PLANT RESPONSES to the ENVIRONMENT

A CRC Series of

Current Topics in Plant Molecular Biology

Peter M. Gresshoff, Editor

Published Titles

Peter M. Gresshoff, PLANT BIOTECHNOLOGY AND DEVELOPMENT, 1992

Peter M. Gresshoff, PLANT RESPONSES TO THE ENVIRONMENT, 1993

Forthcoming

Peter M. Gresshoff, PLANT GENOME ANALYSIS

PLANT RESPONSES to the ENVIRONMENT

Editor

Peter M. Gresshoff
Plant Molecular Genetics
Institute of Agriculture
Center for Legume Research
The University of Tennessee
Knoxville, Tennessee

CRC Press
Boca Raton   Ann Arbor   London   Tokyo

Library of Congress Cataloging-in-Publication Data

Plant responses to the environment / editor, Peter M. Gresshoff.
     p.   cm. -- (A CRC series of current topics in plant molecular
biology)
   Includes bibliographical references and index.
   ISBN 0-8493-8263-7
   1. Botany--Ecology.   2. Plant molecular biology.   I. Gresshoff,
Peter M., 1949-   .   II. Series.
  QK901.P575   1993   75050
  581.5'22--dc20   93-10582
                                                            CIP

Direct all inquiries to CRC Press, Inc., 2000 Corporate Blvd., N.W., Boca Raton, Florida 33431.

© 1993 by CRC Press, Inc.

International Standard Book Number 0-8493-8263-7

Library of Congress Card Number 93-10582

Printed in the United States of America   1 2 3 4 5 6 7 8 9 0

Printed on acid-free paper

THE EDITOR

Peter Michael Gresshoff, Ph.D., D.Sc., holds the endowed Racheff Chair of Plant Molecular Genetics (Institute of Agriculture) at the University of Tennessee in Knoxville.

Professor Gresshoff, a native of Berlin (Germany), graduated in genetics/biochemistry from the University of Alberta in Edmonton in 1970 and then undertook postgraduate studies at the Australian National University in Canberra, Australia, where he obtained his Ph.D. in 1973 and his D.Sc. in 1989.

He completed his postdoctoral work as an Alexander von Humboldt Fellow (1973 to 1975) at the University of Hohenheim (Germany) and Research Fellow (1975 to 1979) in the Genetics Department (RSBS, Australian National University, Canberra), then headed by Professor William Hayes. He was appointed Senior Lecturer of Genetics in the Botany Department at the ANU in 1979, where he built up an internationally known research group investigating the genetics of symbiotic nitrogen fixation. He assumed his present position in Knoxville in January 1988, continuing his research direction by focusing on the macro- and micromolecular changes involved in nodulation. His recent interests have turned to plant genome analysis and DNA fingerprinting.

Professor Gresshoff is also actively involved in the organization of research conferences. In 1990 he chaired the 8th International Congress on Nitrogen Fixation in Knoxville. He also organizes the Annual Gatlinburg Symposia, which focus on current topics in plant biology and molecular genetics. He is the assistant director of the Center for Legume Research.

He was awarded the Alexander von Humboldt Fellowship twice (1973 to 1975 and 1985 to 1986) and is a member of the editorial board of Physiologia Plantarum and the Journal of Plant Physiology. He has received research support from biotechnology firms, competitive grants, the Human Frontier Science Program and the Australian Government. He has published over 165 refereed publications and has contributed to many international and national congresses and symposia. He has been awarded membership in Phi Kappa Phi and Sigma Xi. He is a dedicated teacher and researcher, who believes in technology transfer and innovative science. He has been an advisor on plant biotechnology to the European Commission, the Japanese Government and the Environmental Protection Agency as well as the Department of Energy.

His major research interests concern the characterization of the genes controlling the early establishment of soybean nodules. Focus is given to map-based gene isolation and plant genome analysis as coupled to the cell cycle, signal reception and physiology.

Table of Contents

Preface

This is the second volume in the series *Current Topics in Plant Molecular Biology*. The book focuses on the environment and the way that plants respond to it.

It is not necessary to stress the importance of plants to our life. They provide food and drink, building material, medicines, and narcotics; they are 'clean-air machines', visual screens, beautification in landscape and home, as well as surfaces for our pastime activities like golf, football, or tennis. They clearly affect several nutrient cycles in the biosphere and the weather itself.

Plants also form the basis of the food chain as they convert solar energy to chemical energy. As such they form the basis of almost every industrialized nation. Even for Japan, rice production at home is politically and socially very important. Switzerland would suffer, if her slopes were not green or if forests and pastures were absent.

Agriculture generates income for people and nations. Modern demands of society in terms of productivity of crops and their health value govern activity of plant researchers. Population growth and urbanization have removed significant stretches of prime agricultural lands. Throughout the world, plants are planted under marginal, stressed conditions, leading to reduced productivity and nutritional value. While it sufficed in the past to use chemical alterations of the environment (pesticides, herbicides, fertilizer), these are becoming progressively unpopular because of cost, persistence and environmental impact. To respond to these demands requires an understanding of the ways that genes in plants respond to the environmental factor. It is for this reason that interests in Low Impact Sustainable Agriculture (LISA) and renewable resources for industrial production have increased; this is mirrored by activity in the biotechnology sector.

The first volume of this series concentrated on plant biotechnology and development and provided an overview of techniques and approaches in these areas. This volume presents a palette of different biological systems responding to different sets of biotic and abiotic factors. The term *environment* is taken in its largest context, so that the book contains chapters on both physical and biological factors.

The ability of a plant to respond to the environment is paramount for its adaptability and performance. While many factors were (and are) important for ecological and evolutionary distribution, most present-day interest is fueled by the agricultural significance of plant responses.

Preface

Modern agriculture is affected by environmental factors, both in a beneficial and detrimental way. Plants, being basically sessile, do not escape from a pathogen or environmental stress. They adapt or wither.

This book contains a collection of fascinating chapters, dealing with environmental responses that go beyond the normal physiological stresses. The term *response* is also viewed in a wide way: interaction with pathogens, symbionts, phytohormones and transferred genes, ozone, nitrous oxide and air pollution, microgravity, oxygen deficiency, high temperature and salt stress. Some novel systems that may be used as bioindicators are presented. This explains the inclusion of the chapters on a thermophylic cyanobacterium, *Lotus japonicus*, and soybean transformation. Two chapters deal with the cellular alterations possible after the transfer of exogenous genes. Such gene constructs can be used as reporter cassettes to monitor environmental factors. For example, a cold-sensitive promoter fused to a reporter gene will indicate tissue-specific activation. The application of this technology to environmental toxins and xenobiotics is not far.

Included are also chapters of organisms that are not really plants, although they photosynthesize. These blue green algae or cyanobacteria are unique models for fast analysis of some concepts linking development and environmental stress.

Future volumes in this series will extend the analysis of plant systems. Volume III will focus on *Plant Genome Analysis* and will demonstrate approaches that will aid the classical plant breeder to manipulate crops and predict outcomes of crosses.

As always, I must admit that this volume does not contain all the aspects pertaining to the topic. Oversights were either intentional or accidental. These will be covered by other volumes. For the sake of brevity and a low cost of the book, it is hoped that the present set of chapters gives a broad oversight of the subject and stimulates the reader to pursue the referenced literature. I tried to compile papers from international authors dealing with issues not frequently reviewed in annual reviews. The glossary is added to help the uninitiated reader to manage the jargon of molecular biology and plant biology.

Finally, I want to thank Janice Crockett for valuable help with the production of the book, Michael Gresshoff for help with figures, Brant Bassam with 'Macintoshing' and the Center for Legume Research for hosting the Gatlinburg Symposium, which brought some of the contributing authors together.

Peter M. Gresshoff   February 1993
Knoxville, Tennessee

The Molecular Biology of Systemic Acquired Resistance

Scott Uknes[1], Kay Lawton[1], Eric Ward[1], Thomas Gaffney[1], Leslie Friedrich[1], Danny Alexander[2], Robert Goodman[3], Jean-Pierre Métraux[4], Helmut Kessmann[5], Patricia Ahl Goy[5], Manuela Gut Rella[5] and John Ryals[1]

[1]*Agricultural Biotechnology, CIBA-GEIGY Corporation, P. O. Box 12257, Research Triangle Park, North Carolina 27709;* [2]*Calgene Inc., 1920 Fifth Street, Davis, California 95616;* [3]*Department of Plant Pathology, University of Wisconsin, 1630 Linden Drive, Madison, Wisconsin 53706;* [4] *Institut de Biologie Vegetale et de Phytochimie, Universite de Fribourg 3 rue Albert-Gockel, 1700 Fribourg, Switzerland;* [5]*CIBA-GEIGY Limited, 4002 Basel, Switzerland*

Introduction

Many plant species can be immunized by an inoculation by a necrotizing pathogen. This acquired disease resistance was first documented in the early 1900s and is thought to "play an important role in the preservation of plants in nature" (Chester, 1933). Particularly well-characterized examples of plant immunity are the phenomenon of systemic acquired resistance (SAR) in tobacco (Ross, 1961) and induced resistance in cucumber (Kuc, 1982). In these systems, inoculation by a necrotizing pathogen results in systemic protection against subsequent infections by a number of agronomically important bacterial, fungal and viral pathogens that can last for weeks to months. The important points being that any pathogen which causes leaf necrosis can induce the resistance and the resistance induced by for example a bacterial pathogen can protect the plant against viral, bacterial and fungal pathogens (non-specific immunity). Thus, immunity offers benefits of systemic protection, long lived activity, multiple modes of action, and broad spectrum disease control, all of which are important characteristics for modern disease control strategies.

Our initial experimental approach to understand SAR at a molecular level was based on a working model that could explain the basic steps leading to the establishment of a resistant state. To induce resistance there must first be a necrotic reaction to a pathogen. This requirement for a biological inducer is strict since salt, acid or mechanical damage will not induce SAR. The necrosis triggers the release of a signal molecule, that translocates to various parts of the plant, is perceived, with the final result being the expression of a set of genes responsible for the maintenance of the resistant state. We distinguished the events leading to the onset of resistance (initiation) from events involved in maintaining the resistance (maintenance), although these are probably not completely distinct processes. The goal of this project is to understand SAR at a molecular level and to use this information to genetically engineer crop plants with improved health in the field.

Maintenance of Resistance

Over the past few years we have identified and characterized a number of cDNAs that are expressed to high levels in systemically resistant tissue. Nine of these "SAR gene families" were coordinately expressed in tobacco leaves after treatment with either tobacco mosaic virus (TMV) or 2,6-dichloroisonicotinic acid (INA), a synthetic plant immunomodulator (Ward et al, 1991, Métraux et al, 1991). The timing and the amount of gene expression correlated well with the onset and the degree of resistance to further TMV infection (Ward et al, 1991). The role of each of the genes in mediating disease resistance is currently being evaluated using transgenic expression of the sense and the antisense of the cDNA of each gene.

In order to determine what role a particular gene plays in the maintenance of resistance, each of the SAR cDNAs were subcloned into a high-level expression cassette, based on an enhanced 35S promoter (Kay et al, 1987). The expression cassette containing the cDNA in either a sense or antisense direction was then subcloned into an *Agrobacterium* vector (McBride and Summerfelt, 1990) and leaf disk transformation (Horsch et al, 1985) of Xanthi nc tobacco was carried out. Many independent transformants were selected in each transformation experiment and these were screened for high levels of the protein encoded by the cDNA. The transformants were allowed to self twice and homozygotic lines were established. In each case it was possible to develop genetically stable lines expressing the transgenic protein at levels between 0.1 and 1% of the total protein. When possible the transgenic plants were analyzed for the presence of the transgene, transgene encoded RNA, transgene encoded protein and activity of the transgene encoded protein. Typical protein expression analysis of a transgenic plant containing a acidic class III chitinase from cucumber is shown in Figure 1.

The cucumber protein produced in tobacco is found in the ICF and co-migrates with authentic purified acidic class III cucumber chitinase. In the upper right of Figure 1, the amino terminal amino acid sequence of the acidic class III cucumber chitinase from transgenic tobacco and cucumber is compared after HPLC purification (Métraux et al, 1988). Protein obtained from either source had the same amino terminal sequence indicating faithful translation and processing of the protein in transgenic tobacco. The lower left portion of Figure 1 shows a comparison of chitinase activities of purified acidic class III cucumber chitinase from transgenic tobacco and cucumber. ^{3}H-labeled chitin was incubated with various amounts of purified enzyme as previously described (Métraux and Boller, 1986). Chitinase activity was determined by measuring TCA soluble ^{3}H counts after a 30 minute incubation. Although the enzyme activity varied from experiment to experiment enzyme purified from either source was virtually identical.

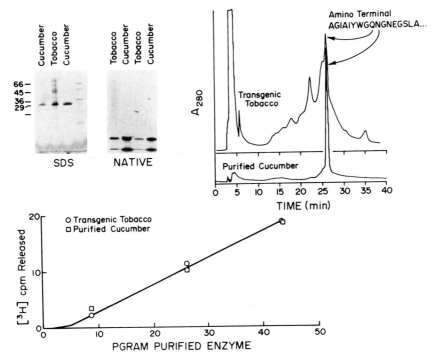

Figure 1. *Characterization of a transgenic tobacco plant which expresses an acidic class III cucumber chitinase. Upper left. Denaturing and native gel analysis of purified acidic class III cucumber chitinase (cucumber) and intercellular wash fluid (ICF) from a transgenic tobacco plant expressing the acidic class III cucumber chitinase (tobacco).*

The lines were then evaluated for resistance against a number of viral, fungal and bacterial diseases as well as several insects. Examples of resistance with single genes against the following three diseases have been observed: 1) post-emergent damping off (*Rhizoctonia solani*) in transgenic tobacco expressing high levels of either the basic class I chitinase, the acidic class III chitinase or the cucumber class III chitinase; 2) black shank (*Phytophthora parasitica*) in transgenic tobacco expressing high levels of the protein SAR8.2, a tobacco SAR protein with unknown structural or enzymatic function; 3) bluemold (*Peronospora tabacina*) in transgenic tobacco expressing high levels of pathogenesis-related protein 1 (PR-1). This resistance to bluemold by PR-1 has been demonstrated to be statistically significant in multiple independent lines (Figure 2).

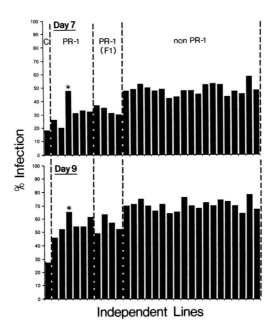

Independent Lines

Figure 2. Resistance to Peronospora tabacina *in transgenic tobacco expressing PR-1 protein constitutively. Ten plants for each line were artificially inoculated with* Peronospora tabacina *in a greenhouse. Plants were assayed at 7 and 9 days after inoculation by measuring the percentage of leaf area showing disease symptoms. The average for each line is shown. Line labeled 'C' is a immunization compound treated control. Lines labeled PR-1 are plants homozygous for the 35S-PR-1 transgene. However, the line marked with an * does not express any PR-1 mRNA or protein constitutively. Lines marked PR-1 (F1) are heterozygous for the 35S-PR-1 transgene. Lines labeled non PR-1 are transformants that do not contain the 35S-PR-1 transgene.*

At seven days after infection the PR-1a expressing transgenic lines have a disease reduction of 42% relative to the non-PR-1a controls. However, at day 9 the reduction of disease is 22% compared to the controls, possibly indicating the involvement of other gene products for full immunization. The reductions in disease at both day seven and nine are statistically significant (pairwise t-test) at > 99% confidence. Therefore, it appears that PR-1 in tobacco effectively functions as an antifungal protein, specifically causing some level of resistance to *Peronospora tabacina*. Other factors, perhaps other SAR gene products probably contribute to the full immunity observed in the chemically treated control.

Crosses of each homozygotic line to a tobacco line carrying a different transgene have also been made and these lines have been evaluated. There are several examples of resistance when two genes are expressed in a line that were not observed in lines expressing either single gene. In addition, pairs of SAR genes have been engineered into the same vector. These lines have been crossed to give single plants expressing three or four SAR genes at high levels.

The results of these experiments has led us to a working model for the activity of the genes expressed in SAR. Several of these SAR genes may target one particular pathogen and different, overlapping sets of SAR genes may affect different pathogens. It will require a systematic and detailed study of the interactions of the different SAR genes in relation to various pathogens in order to determine which set of genes to use to give optimal protection against a specific disease.

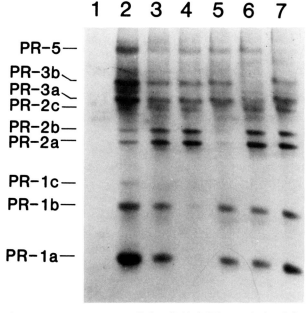

Figure 3. Analysis of antisense plants. Intercellular fluid (ICF) was isolated from tobacco 7 days after inoculation with tobacco mosaic virus (TMV, lanes 2-7) or buffer control (lane 1) and analyzed by native polyacrylamide gel electrophoresis (Coomassie blue stained). Lane 1, primary tissue, Xanthi; Lane 2, primary tissue, Xanthi; Lane 3, secondary tissue, Xanthi; Lane 4, secondary tissue, PR-1 antisense plant; Lane 5, secondary tissue, PR-2 antisense plant; Lane 6, secondary tissue, PR-3 antisense plant; Lane 7, secondary tissue, PR-5 antisense plant.

We have also taken the approach of expressing high levels of the antisense RNA to each SAR gene (Figure 3, PR-4 not shown). Homozygotic transgenic lines were developed for each of the genes and the protein target is dramatically reduced in each case. Interestingly, when anti-PR-1a is expressed all three PR-1 proteins (a, b and c) are dramatically reduced in abundance (Figure 3). When antisense to PR-2a is made PR-2a, b and c are all reduced. When antisense to PR-3b is made PR-3 a, and b are reduced. When antisense to PR-5 is expressed PR-5 protein accumulation is inhibited. This indicates that homologous as well as identical gene products are effectively eliminated by antisense directed against only one member of a gene family. Two further vectors have been constructed, one that contains anti-PR-1, anti-PR-2 and anti-PR-4 and one that contains anti-PR-3 and anti-PR-5.

The Signal for SAR

In previous work we demonstrated the appearance of a fluorescent compound in phloem exudate of cucumber plants that had been induced to resistance by either tobacco necrosis virus or *Colletotrichum lagenarium*. The compound was purified to homogeneity and determined to be salicylic acid (SA) using gas chromatography/mass spectral analysis (Métraux et al, 1990). SA accumulated to high levels prior to the onset of induced resistance to *C. lagenarium* making it a candidate for the signal molecule. In independent experiments, Malamy et al. (1990) demonstrated that SA was induced in tobacco that had been induced with TMV and Hammerschmidt's group demonstrated the induction of SA in cucumber preceding the development of resistance induced by *Pseudomonas syringae* (Rasmussen et al, 1991). If SA were the signal molecule for SAR in tobacco then exogenously applied SA should coordinately induce the same SAR genes as are found in the biologically induced resistance. Ward et al (1991) demonstrated that SA coordinately induced the same nine gene families as TMV in tobacco and that the timing and extent of expression correlated with the onset and degree of protection against further TMV infection (Ward et al, 1991). Therefore, there is considerable evidence implicating SA as a signal molecule involved in triggering resistance.

When the biosynthetic pathway of SA is considered the finding of SA as a putative signal becomes particularly interesting. Of the two proposed pathways for SA biosynthesis in higher plants, both branch from the phenylpropanoid pathway after the synthesis of trans-cinnamic acid (Figure 4).

It has been known for a long time that the phenylpropanoid pathway is induced following pathogen attack and products of this pathway, such as isoflavonoid phytoalexins and lignin precursors, are used in disease resistance. It is clear that SA production could be a further consequence of inducing this pathway and this could provide an answer as to why a biological inducer is required for signal production.

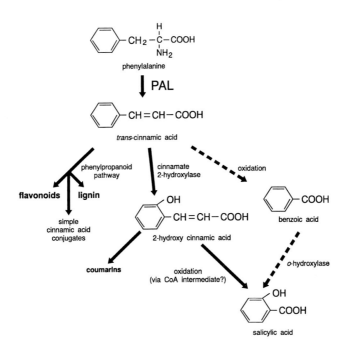

Figure 4. Possible biosynthetic pathway for SA (reprinted with permission from The Plant Cell 3,1085-1094).

As appealing as this model may be, there are two results that cast some doubt on the role of SA as the signal molecule. First, it is clear from our studies and from work of others that exogenously applied SA does not translocate well in the plant. Of course, it can be argued that the application of a compound to a leaf surface is not tantamount to the production and secretion of the compound *in vivo*. A second interesting result came for Hammerschmidt's studies with cucumber (Rasmussen et al, 1991). They found that cucumber leaves infiltrated with a suspension of *Pseudomonas syringae* do not accumulate high levels of SA within four hours after infiltration. However, removal of the leaves at four hours post-infiltration does not prevent the accumulation of high levels of SA or the onset of resistance in other parts of the plant (Rasmussen et al, 1991). It can be argued that SA induces its own biosynthesis, which is indeed similar to the biosynthesis of ethylene (Yang and Hoffman, 1984) or that the level of detection is not sensitive enough. However, the presence of a signal which induces SA production can not be ruled out. Therefore, it is still not clear that SA is the translocated signal that induces SAR.

In order to investigate the role of SA as a signal molecule for SAR we have engineered plants that degrade salicylate. It is known that certain strains of *Pseudomonas putida* are capable of degrading naphthalene through salicylate to acetaldehyde and pyruvate. One step of this pathway is the conversion of salicylate to catechol, which we have shown does not induce SAR. This conversion is carried out by salicylate hydroxylase an enzyme encoded by the NahG gene. Therefore, we have cloned the NahG coding sequence into different plant expression vectors and developed transgenic tobacco expressing high levels of the corresponding mRNA. In preliminary experiments we have found that plants that express high levels of the mRNA are not capable of inducing SAR, while plants expressing moderate levels give moderate levels of resistance and transformants with undetectable levels of the mRNA respond normally (Figure 5).

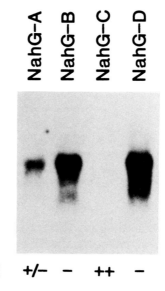

SAR   +/-   –   ++   –

Figure 5. Effect of expression of salicylate hydroxylase (NahG) on SAR in tobacco. Four independent transgenic tobacco containing 35S-NahG were analyzed for both NahG expression (RNA gel blot analysis) and SAR (reduction of TMV lesion number and size). Thus the data support the role of SA as a signal molecule that induces the SAR genes which then provide the observed resistance.

Summary

The molecular events that take place during the initiation and maintenance of SAR are starting to be resolved. Based on the results presented in this report, our current working model is that the infection by a pathogen induces the synthesis of salicylic acid which translocates throughout the plant and is recognized by a specific receptor. The receptor transduces the signal with the result being the coordinate expression of a set of SAR genes. The encoded SAR proteins accumulate to high levels systemically. The resistance observed is provided by certain groups of proteins acting on particular pathogens.

References

Chester, K. (1933) *Quarterly Reviews of Biology* 8, 275-234.

Horsch, R., Fry, J., Hoffmann, N., Eichholtz, D., Rogers, S. & Fraley, R. (1985) *Science* **227**, 1229-1231.

Kay, R., Chan, A., Daly, M. & McPherson, J. (1987) *Science* **236**, 1299-1302.

Kuc, J. (1982) *Bioscience* **32**, 854-860.

Malamy, J., Carr, J., Klessig, D. & Raskin, I. (1990) *Science* **250**, 1002-1004.

McBride, K. & Summerfelt, K. (1990) Plant Molecular Biology **14**, 269-276. Métraux, J.-P. and Boller, T. (1986).*Physiol. Mol. Plant Pathol.* **56**, 161-169.

Métraux, J.-P., Goy, P., Staub, T., Speich, J., Steinemann, A., Ryals, J. & Ward, E. (1991) In: *Advances in Molecular Genetics of Plant Microbe Interactions.* Hennecke, H. and Verma, D. (eds). Vol. 1. 432-439. Kluwer Press, Dordrecht.

Métraux, J.-P., Signer, H., Ryals, J., Ward, E., Wyss-Benz, M., Gaudin, J., Raschdorf, K., Schmid, E., Blum, W. & Inveradi, B. (1990) *Science* **250**, 1004-1006.

Métraux, J.-P., Streit, L. & Staub, T. (1988) *Physiol. and Mol. Plant Path.* **33**, 1-9.

Rasmussen, J., Hammerschmidt, R. & Zook, M. (1991) *Plant Physiology* **97**, 13424-1347.

Ross, A. (1961) *Virology* **14**, 340-358.

van Loon, L. (1985) *Plant Molecular Biology* **4**, 111-116.

Ward, E., Uknes, S., Williams, S., Dincher, S., Wiederhold, D., Alexander, D., Goy, P., Métraux, J.-P. & Ryals, J. (1991) *Plant Cell* **3**, 1085-1094.

Yang, S. & Hoffman, N. (1984) *Ann. Rev. Plant Physiol.* **35**, 155-189.

Alcohol Dehydrogenase (*Adh*) Genes and their Expression in Higher Plants: Maize and Petunia

Judith Strommer¥, Robert G. Gregerson*, Elizabeth Foster¥ and Shurong Huang°

¥*Departments of Horticultural Science and Molecular Biology and Genetics, University of Guelph, Guelph, Ontario N1G 2W1, Canada; * Department of Agronomy and Plant Genetics, University of Minnesota, St. Paul, MN 55108, USA; ° Department of Genetics, University of Georgia, Athens, GA 30602*

General patterns of ADH production in higher plants

Built on the discovery of alcohol dehydrogenase (ADH) induction in response to oxygen deprivation in maize seedlings (Hageman and Flesher, 1960) and the development of maize *Adh* as a genetic system (Schwartz and Endo, 1966; Freeling, 1973; Freeling and Schwartz, 1973), ADH isozymes of maize and the genes encoding them became standards for research on the molecular biology of higher plants. Electrophoretic gels stained for ADH activity revealed the presence of one ADH isozyme in pollen and developing seeds and three in oxygen-deprived roots (Freeling and Schwartz, 1973). These three products result from the differentially charged products of two genes, *Adh1* and *Adh2*, which form active homodimers and heterodimers, the latter with intermediate electrophoretic mobility. While the role of ADH enzymatic activity in pollen and seed is not obvious, its known function in anaerobic glycolysis provided an early rationale for anaerobic induction (Hageman and Flesher, 1960). Early studies focused on maize roots; Okimoto et al (1980) subsequently verified the production of ADH in several maize tissues under hypoxic conditions.

In maize ADH has been shown to accumulate in response to a variety of treatments including low temperature (Christie et al, 1991), application of 2,4-D (Freeling, 1973) and abscisic acid (Hwang and VanToai, 1991). The response to 2,4-D is exhibited by rice (Ricard et al, 1986); in potato, elicitor treatment and wounding also result in increased levels of ADH (Butler et al, 1990). In general patterns of ADH activity are well conserved among higher plants, including tomato (Tanksley et al, 1981), rice (Ricard et al, 1986), barley (Good and Crosby, 1989), and soybean (Tihanyi et al, 1989; Russell et al, 1990).

The induction of ADH associated with oxygen deprivation, generally described as anaerobic induction in the literature, may be more precisely considered to occur in response to hypoxia in terms of both its stimulus in nature and its experimental manipulation. "Anaerobic" and "hypoxic" will both be used in this article, depending on the context. Both refer to effects of oxygen deprivation, resulting in the "anaerobic response" described by Sachs et al (1980).

The *Adh* genes of maize were early cloning targets. The inducibility of ADH activity and associated increase in translatable ADH-RNA (Ferl and Schwartz, 1980) afforded the hope of recovering cDNA clones with relative ease. The presumed transcriptional induction of ADH-RNA during hypoxic stress was supported by the efficient recovery of clones and subsequent Northern hybridization data (Gerlach et al, 1982; Dennis et al, 1985), and was later confirmed by direct measurement of transcription (Rowland and Strommer, 1985). In maize roots, the increase in ADH-RNA is highest near the root tip, primarily in cortical and epidermal cells (Rowland et al, 1989). *Cis-* and *trans-* acting elements associated with induction of *Adh1* or *Adh2* have now been identified (Ferl, 1990; Paul and Ferl, 1991).

Comparison of transcriptional induction and steady-state levels of ADH1-RNA during stress and recovery of maize roots has revealed that stability of the stress-induced message depends on oxygen availability: The message is stable, with a half-life of approximately 16 hours, under anaerobic conditions but quickly drops several-fold with the reintroduction of air (Rowland and Strommer, 1986). The fast turnover of induced messages with reintroduction of oxygen is a feature which appears to be shared by several hypoxically induced RNA species (Hake et al, 1985).

The selective turnover of stress-induced messages may be linked to their preferential translation under conditions of hypoxic stress. Nearly a quarter-century ago Lin and Key (1967) demonstrated a rapid decrease in large polysomes, and a concomitant increase in monosomes and small polysomes, after subjecting soybean rootlets to anaerobic conditions. Comparison of the few hypoxically-induced root polypeptides synthesized *in vivo* and the many translatable *in vitro* suggested selective translation of stress-induced messages in maize (Sachs et al, 1980). These studies relied on the labeling of intact tissues, leaving open the possibility that uptake or distribution of radioactive precursors, together with the relative abundance of stress-induced messages, might result in an apparent but artifactual cessation of normal protein synthesis. The work of Bailey-Serres and Freeling (1990) has provided direct evidence for an effect of hypoxia on translation. They demonstrated a change in the polysomes of maize such as that reported by Lin and Key, and related it to hypo-phosphorylation of a 31 kD ribosomal protein. Similar results have been obtained by Crosby and Vayda with stressed potato tubers (1991).

Of particular relevance to this study is the differential expression of the *Adh* genes of maize. In summary, ADH1 is produced in pollen grains and scutellum; ADH1 and ADH2 are both expressed to low levels in many vegetative tissues and are induced several-fold in hypoxic tissues. Their stress-related induction appears to be regulated at the levels of transcription, messenger utilization and RNA turnover. Although best characterized in maize, hypoxic induction and differential expression of Adh genes comprising a small gene family appears to be a general characteristic of higher plants (Tanskley, 1979; Tanskley and Jones, 1981; Good and Crosby, 1989; Xie and Wu, 1989), excluding Arabidopsis, the single known example of a genus with one *Adh* locus (Cheng and Meyerowitz, 1986).

We decided a few years ago that a thorough characterization of *Adh* genes, their expression, and their regulation in a plant distant from maize would help explain the rationale for ADH1 and ADH2 production under various conditions and, in a larger context, shed light on the evolution and maintenance of small gene families in higher plants. We sought a dicot species with well-developed genetics, transformation capability and horticultural significance. The choice was a tropical woody perennial known in cooler regions as an herbaceous annual bedding plant, *Petunia hybrida*.

Cloning the *Adh* genes of petunia

When plantlets of petunia are submerged in water or, alternatively, set into a chamber supplying argon or nitrogen in place of air, oxygen is slowly depleted. In response, the plants produce elevated levels of three ADH isozymes separable on electrophoretic gels (Figure 1). The pattern is very similar to that of maize, suggesting the induction of two genes encoding polypeptides of different electrophoretic mobility, active as homodimers and heterodimers.

The first petunia *Adh* gene to be cloned was recovered from a lambda-based genomic library by virtue of its homology to a maize cDNA probe (Gregerson et al, 1991). Its intron-exon structure, deduced from DNA sequencing and mapping of the start site, is depicted in Figure 2. The structure is that typical of plant *Adh* genes, with ten exons and nine introns, distinguishable solely by the unusually great length of introns. Correlation of this gene with the slower migrating ADH polypeptide was accomplished by transfecting petunia protoplasts with a copy of this gene fused to a cauliflower mosaic virus 35S promoter and demonstrating the consequent appearance of a slow-

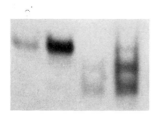

Figure 1. Acrylamide gels stained for ADH activity show the predominance of one isozyme in seeds (lane 1) and mature anthers (lane 2). Non-stressed roots of young plants (lane 3) contain low levels of three isozymes which are induced proportionately by hypoxia (lane 4). The pattern for stems and leaves is the same as that for roots. Plants used for analysis were six weeks old; induction was for 14 hours Induction and ADH analysis were as described in Gregerson et al (1991).

1   2   3   4

migrating band of ADH homodimers (Gregerson et al, 1991). We designated this gene *Adh1*.

Using a maize cDNA probe and hybridizing Southern blots at moderate stringency one finds three fragments of petunia DNA, one of which matches *Adh1* (Figures 3A and 3B). We sought without success to clone the remaining *Adh*-like sequences using total of three genomic libraries from which a dozen copies of *Adh1* were recovered. As an alternative we used degenerate primers to amplify genomic DNA sequences stretching between conserved regions of exon 2 and exon 8 (Gregerson et al, 1993). Two bands of amplification products were detectable in agarose gels. Both were excised for cloning and four clones of each, from independent amplifications, were sequenced for several hundred bases pairs, verifying that each band represented one fragment of amplified DNA. One product was identical to *Adh1*; the other has been designated *Adh2*. Its structure is shown in Figure 2, and its Southern blot hybridization pattern in Figure 3C.

The third gene predicted from Southern blot analysis has not been recovered. From its reproducible appearance in Southern blots using the inbred variety V30 we are confident it is not an allele of *Adh1* or *Adh2*. It may represent an *Adh*--like gene without sufficient conservation of exons 2 &/or 8 to permit amplification. We have no explanation for our failure to recover either this gene or *Adh2* from genomic libraries, except that we have had difficulty in cloning the amplified 5'-proximal region of *Adh2*. It is possible that the sequences involved are not maintained in *E. coli* or that they confer some disadvantage to the host.

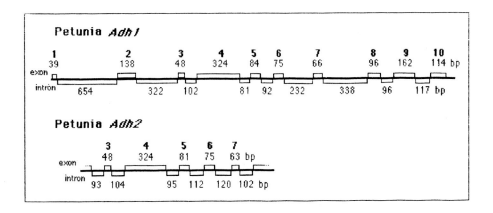

Figure 2. Exon/intron structures of Petunia Adh genes deduced from DNA sequence data.

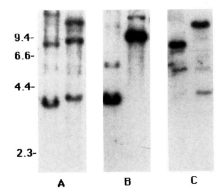

Figure 3. Southern blot of Petunia hybrida V30 DNA probed first with maize ADH1-cDNA (A), then petunia Adh1 (B), and petunia Adh2 (C). Ten micrograms of DNA were cut with either BglII (left lanes) or EcoRI(right lanes). Positions of markers are indicated in kbp.

The amino acid sequences deduced for the four identified *Adh* genes of maize and petunia are presented (Fig. 4). Comparing the regions for which petunia ADH2 sequence is available, one sees that conserved regions are roughly clustered for the four polypeptides. The two petunia genes share 85.7% (90.8%) identity, where the numbers in parentheses take into account both identity and conservative replacements. Overall, petunia ADH1 has 87.3% (92.3%) sequence identity to maize ADH1 and 84.3% (89.9%) identity to maize ADH2. For petunia ADH2, the corresponding numbers are 85.6% (92.8%) and 85.1% (91.1%). Thus, although they share one insertion/deletion with maize ADH2, both petunia genes are more like maize ADH1 for the known region. The greatest identity is between the two ADH1 polypeptides.

```
Maize    ADH1  MA..TAGKVIKCK   AAVAWEAGKPLSIEEVEVAPPQAMEVRVKILFTSL
         ADH2  MA..TAGKVIKCR   AAVTWEAGKPLSIEEVEVAPPQAMEVRIKILYTAL
Petunia  ADH1  MSSNTAGQVIRCK   AAVAWEAGKPLVIEEVEVAPPQKMEVRLKILFTSL
         ADH2  --------------------------------------------------


Maize    ADH1  CHTDVYFWEAK   GQTPVFPRIFGHEAGG   IIESVGEGVTDVAPGDHVL
         ADH2  CHTDVYFWEAK   GQTPVFPRILGHEAGG   IVESVGEGVTDVAPGDHVL
Petunia  ADH1  CHTDVYFWEAK   GQTPLFPRIFGHEAGG   IVESVGEGVTDLKPGDHVL
         ADH2  ---DVYFWEAK   GQNPVFPRILGHEAAG   IVESVGEGVTELVPGDHVL


                                                         G
Maize    ADH1  PVFTGECKECAHCKSAESNMCDLLRINTDRGVMIADGKSRFSINGKPIYH
         ADH2  PVFTGECKECAHCKSEESNMCDLLRINVDRGVMIGDGKSRFTISGQPIFH
Petunia  ADH1  PVFTGECQQCRHCKSEESNMCDLLRINTDRGVMIHDGQTRFSKDGKPIYH
         ADH2  PVFTGECKDCAHCKSEESNMCSLLRINTDRGVMIHDGQSRFSINGKPIFH


Maize    ADH1  FVGTSTFSEYTVMHVGCVAKINPQAPLDKVCVLSCGIST   GLGASINVA
         ADH2  FVGTSTFSEYTVIHVGCLAKINPEAPLDKVCILSCGIST   GLGATLNVA
Petunia  ADH1  FVGTSTFSEYTVCHSGCVTKIDPQAPLDKVCVLSCGIST   GLGATLNVA
         ADH2  FVGTSTFSEYTVVHVGCLAKINPLAPLDKVCVLSCGIST   GLGATL.VA


Maize    ADH1  KPPKGSTVAVFLGLGAVGLA   AAEGARIAGASRIIGVDLNPSRFEE   A
         ADH2  KPAKGSTVAIF.GLGAVGLA   AMEGARLAGASRIIGVDINPAKYEQ   A
Petunia  ADH1  KPTKGSTVAIF.GLGAVGLA   AAEGARIAGASRIIGVDLNPSRFND   A
         ADH2  KPTKGSSVAIF.GLGAVGLA   AAEGARIAGASRIIDVDLNASRFEE   A


Maize    ADH1  RKFGCTEFVNPKDHN.KPVQE   VLAEMTNGGVDRSVECTGNINAMIQAF
         ADH2  KKFGCTEFVNPKDHD.KPVQE   VLIELTNGGVDRSVECTGNVNAMISAF
Petunia  ADH1  KKFGVTEFVNPKDHGDKPVQQ   VIAEMTDGGVDRSVECTGNVNAMISAF
         ADH2  KKFGVTEFVNPKDYS.KPVQR   VIAEMTDGGVNRSVECTGHIDAMISAF


Maize    ADH1  ECVHD   GWGVAVLVGVPHKDAEFKTHPMNFLNERTLKGTFFGNYKPRTD
         ADH2  ECVHD   GWGVAVLVGVPHKDDQFKTHPMNFLSEKTLKGTFFGNYKPRTD
Petunia  ADH1  ECVHD   GWGVAVLVGVPNKDDAFKTHPMNLLNERTLKGTFFGNYKPKSD
         ADH2  ECVHD------------------------------------------------


                                               D
Maize    ADH1  LPNVVELYMKK   ELEVEKFITHSVPFAEINKAFNLMAKGEGIRCIIRMEN
         ADH2  LPNVVEMYMKK   ELELEKFITHSVPFSEINTAFDLMLKGEGLRCIMRMED
Petunia  ADH1  IPSVVDKYMKK   ELELEKFITHQVPFSEINKAFDYMLKGESIRCMITMEH
```

Figure 4. Deduced amino acid sequences for the ADH polypeptides of maize and petunia. Maize ADH sequences are as reported by Dennis et al (1985); petunia ADH sequences from Gregerson et al (1991; 1993). Exon/intron boundaries are indicated by double spaces. Known differences between maize Adh alleles are indicated with alternate peptides above the Maize ADH1 line.

As expected, however, a nearest-neighbor joining tree, which includes all the plant *Adh* sequences available at the time, separates the genes of petunia from those of maize, with monocots on one side and dicots on the other (Gregerson et al, 1993).

Expression of *Adh1* and *Adh2* in petunia

With histochemical staining one can examine the tissue-specific pattern of ADH activity. Micrographs of mature pollen grains, young primary roots, and young primary roots hypoxically induced by submersion in water are presented in Figure 5. A high level of ADH activity is apparent in pollen grains (5A). This activity is the result of *Adh* gene expression in maturing pollen grains, beginning when floral buds are between 4 and 5 cm long (Gregerson et al, 1991). Leaves and roots contain low but often detectable levels of ADH activity, visible as bits of dark precipitate seen particularly near the root stele (5B). We believe this ADH reflects a constitutive state of hypoxia

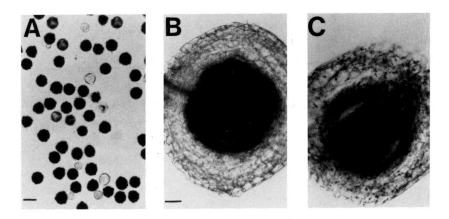

Figure 5. Histochemical staining of mature pollen (A), and primary roots of aerobically growing(B), and hypoxic (C) petunia plantlets. ADH activity results in deep staining of pollen grains and dark precipitate in root cells. Darkness of the stele is normal and does not reflect ADH activity. The bar in panel A represents 150 nm; that in panel B, 500 nm.

in some cells due to air diffusion barriers and, in the case of roots, to proximity of the stele, which serves as an oxygen sink. Hypoxia results in greatly elevated levels of ADH activity throughout the living tissues of the root, as shown in panel C. In leaves, although we routinely see patches of uninduced cells (data not shown), all areas of the leaf appear susceptible to ADH induction.

In contrast to histochemical staining, which permits identification of cell types producing ADH, the staining of electrophoretic gels containing tissue extracts allows us to measure the relative amount of activity attributable to ADH1 and ADH2 gene products. Such an activity gel is depicted in Figure 1, from which it can be seen that anther contains a great deal of ADH, all apparently ADH1. This is due to the high accumulation of ADH1 in pollen; non-germinal anther tissue expresses ADH1 and ADH2 to about the same extent as vegetative tissues (data not shown). The ADH activity of seeds is also predominantly ADH1. All other tissues examined, including aerobically grown roots, leaves and stems, express a low level of both gene products with a predominance of ADH2. The levels of ADH in aerobic tissues varies, apparently in relation to cycles of watering. With induction the ADH activity associated with both polypeptides increases, with ADH2 accounting for approximately 70% of total activity.

DNA sequences of petunia *Adh1* and *Adh2* are sufficiently different that they can be used as gene-specific probes in Northern hybridizations (Figure 6). As predicted from activity gels, ADH1-RNA is inducible and present in anther (the anther lane contains 1/10 the amount of RNA loaded in the other two lanes); ADH2-RNA is detectable in untreated plantlets and induced by hypoxia.

We conclude from these analyses that living cells of most or all tissues express low fluctuating levels of ADH, which increase significantly with hypoxic stress. ADH1 and ADH2 polypeptides both contribute to the ADH activity seen in vegetative tissues, with a larger contribution from ADH2.

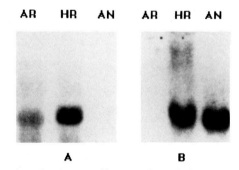

Figure 6. Northern blot hybridization of petunia RNA isolated from aerobic roots (AR), hypoxic roots (HR), and mature anthers (AN) probed with the petunia Adh2 amplification product (panel A) or a fragment isolated from exon 4 of petunia Adh1 (panel B). Root lanes contain ten times more RNA than the anther lane.

ADH1 alone is present at high levels in developing pollen grains. Leaves are not immune from induction. Although the absence of an anaerobic response has been reported for maize leaves (Okimoto et al, 1980), it is characteristic of induced leaves of rice (Xie and Wu, 1989), and we find elevated levels of anaerobic peptides translatable *in vitro* from induced maize leaf RNA (Huang, unpublished data).

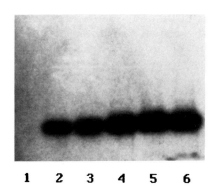

Figure 7. Northern blot hybridization of petunia root polyadenylated RNA isolated from plantlets induced hypoxically for 0, 2, 4, 8, or 16 hours (lanes 1-5 respectively, 2 μg RNA) and from mature anthers (lane 6, 0.2 μg RNA). The blot was probed at high stringency with a petunia Adh1 fragment isolated from exon 1 of the cloned gene.

Perhaps these induced messages in maize leaves are inefficiently translated *in vivo*, but in terms of transcript abundance the anaerobic or hypoxic response occurs in maize leaves as well as roots. At the level of gene expression, then, the tissue- and stimulus-specific recruitment of members of the *Adh* gene family has been well conserved in evolution.

Regulation of *Adh* gene expression in petunia

The differential pattern of enzymatic activity for ADH1 and ADH2 parallels the pattern of mRNA accumulation, and in large measure reflects differential patterns of gene expression. Northern analysis of ADH1-RNA reveals that mRNA accumulates in roughly linear fashion for approximately eight hours, and the maximal level is maintained for several hours (Figure 7). The rapid increase would be difficult to explain by changes in the pattern of RNA turnover; we therefore conclude that induction involves increased transcription. Again, the pattern is very similar to that of maize and other plants.

With radioactive labeling of nascent polypeptides *in vivo* we see that a significant decrease in "normal," aerobically synthesized products accompanies anaerobic induction, while products of hypoxia-induced messages are readily visualized (S. Huang, unpublished data). Selective translation of stress-induced messages is a thorny problem to address by these means, however, as discussed in the first section of this chapter. Until we have analyzed polysomes or cultured cells to eliminate physiological parameters we cannot verify the selective translation of stress-induced messages in petunia. Comparing the results of Sachs et al (1980) with corn to ours from petunia, however, the changes in polypeptide synthesis revealed by radioactive labeling of amino acids *in vivo* are very similar.

The stress-dependent stability of hypoxia-induced messages is also difficult to assess quantitatively in plants (Rowland and Strommer, 1986), but the quick turnover of induced messages accompanying a return of petunia plantlets to aerobic conditions is revealed by the pattern of polypeptides translated *in vitro*, as shown in Figure 8. Over a period of 15 hours roots accumulate high levels of several translatable RNA species, including that of petunia ADH (40 kD, Gregerson et al, 1991). Within two hours of the reintroduction of oxygen, the level of this RNA has returned to that of unstressed controls (lane 4). The means by which stress-induced ADH messages are identified for aerobic degradation is not known. We have determined that the polyadenylate tails of stress-induced ADH-RNAs are indistinguishable from those of aerobic RNAs (Macdowell and Strommer, unpublished data), eliminating the most easily tested hypothesis for recognition of stress-induced transcripts. Under-phosphorylation of a stress-associated ribosomal protein in maize (Bailey-Serres and Freeling, 1991) may provide a clue to explain the observed selective degradation.

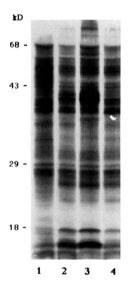

Figure 8. Translation of polyadenylated petunia RNA in vitro. RNA was isolated from aerobic roots (lane 1), from roots hypoxically induced for seven (lane 2) and 15 (lane 3) hours and from roots induced for 15 hours then returned to an aerobic environment for two hours (lane 4). Positions of size markers are indicated on the left; ADH monomers are approximately 40 kilodaltons.

Summary and Conclusions

In both maize and petunia there are two identifiable genes encoding ADH polypeptides; in both species these genes are expressed differentially, a pattern typical of higher plants. We find no significant differences in the way the genes are utilized in the two species: *Adh1* is induced by hypoxia and expressed developmentally in seed and pollen; *Adh2* appears to be expressed only in response to hypoxia. Analysis of protein binding sites in the promoters of *Adh1* and *Adh2* of maize (Ferl, 1990; Paul and Ferl, 1991) provides experimental evidence for basic differences in the modes of regulation operating on the two genes. These results suggest that promoters for *Adh1* and *Adh2* respond to different or overlapping sets of signals. One promoter or promoter region, for example, might respond directly to changes in oxygen pressure while another might respond to low pH associated with early stages of anaerobic glycolysis. In both species the anaerobic response involves regulation at the transcriptional level, accompanied with regulation at the level of RNA turnover and probably RNA utilization as well.

The evolutionary maintenance of both sets of strategies -- circumstance-specific activation of specific genes and significant reliance on both transcriptional and post-transcriptional regulation -- in such distant species argues for the success of these strategies in meeting the challenges presented by an unstable, inescapable environment. On one hand, the use of distinct methods for activating two genes of very similar function allows for their individual recruitment under different sets of conditions, for example, cyclic hypoxia and microspore development. On the other hand, the regulation of mRNA availability (and probably utilization) allows for a quick, appropriate response to a transient and fluctuating stress.

Our early results suggest many research paths. We would like to examine the utilization and turnover of ADH-RNA during pollen maturation, for levels appear to drop precipitously at a specific time in pollen maturation (Gregerson, unpublished data). We plan to look carefully at the translation of stress- and nonstress-induced messenger RNA in cultured cells to avoid whole-plant associated problems of precursor uptake and distribution. A very promising avenue of research involves analysis of the *Adh* promoters in petunia as a means of determining the difference in routes of hypoxia-dependent gene activation. Petunia provides an especially user-friendly system for all these studies. It produces an abundance of seed, adapts readily to culture, is susceptible to *Agrobacterium*-mediated transformation, and regenerates well from culture. What we learn from analysis of *Adh* genes and their deployment in petunia promises to apply to many plants carrying *Adh* as a small gene family, providing answers to questions related to the physiology, regulation and evolution of ADH and its genes in higher plants.

Acknowledgments

The research reported here was funded by NSERC operating grant OGPOO46636 and, in its early stages, by NIH grant GM38616 to JS. RG was supported by an NIH training grant and EF by an NSERC postgraduate scholarship.

References

Bailey-Serres, J. & Freeling, M. (1990) *Plant Physiology* **94**, 1237-1243.

Cheng, C. & Meyerowitz, E.M. (1986) *Proc. Natl. Acad. Sci. USA* **83**, 1408-1412.

Christie, P.J., Hahn, M. & Walbot, V. (1991) *Plant Physiology* **95**, 699-706.

Crosby, J.S. & Vayda, M.E. (1991) *Plant Cell* **3**, 1013-1023.

Dennis, E.S., Sachs, M., Gerlach, W., Finnegan, E. & Peacock, W. (1985) *Nucleic Acids Research* **13**, 727-743.

Ferl, R.J. (1990) *Plant Physiology* **93**, 1094-1101.

Ferl, R.J., Brennan, M.D. & Schwartz, D. (1980) *Biochemical Genetics* **18**, 681-691.

Freeling, M. (1973) *Molecular and General Genetics* **127,** 215-227.

Freeling, M. & Schwartz, D. (1973) *Biochemical Genetics* **8**: 27-36.

Gerlach, W.L., Pryor, A., Dennis, E., Ferl, R., Sachs, M.M. & Peacock, W. J. (1982) *Proc. Natl . Acad. Sci. (USA)* **79**, 2981-2985.

Good, A.G. & Crosby, W.L (1989) *Plant Physiology* **90**, 860-866.

Gregerson, R.G., McLean, M.M., Beld, M., Gerats, A.G.M. & Strommer, J. (1991) *Plant Molecular Biology* **17**, 37-48.

Gregerson, R.G., Cameron, L., McLean, M., Dennis, P., & Strommer, J. (1993) *Genetics* (in press)

Hageman, R.H. & Flesher, D. (1960) *Archives of Biochemistry and Biophsics* **87**, 203-209.

Hake, S., Kelley, P.M., Taylor, W.C. & Freeling, M. (1984) *J. Biol. Chem.* **260**, 5050-5054.

Hwang, S.-Y. & VanToai, T.T. (1991) *Plant Physiology* **97**, 593-597.

Lin, C.Y. & Key, J.L. (1967) *Journal of Molecular Biology* **26**, 237-247.

Okimoto, R., Sachs, M.M., Porter, E.K. & Freeling, M. (1980) *Planta* **150**, 89-94.

Paul, A.-L. & Ferl, R.J. (1991) *Plant Cell* **3**, 159-168.

Ricard, B., Mocquot, B., Fournier, A., Delseny, M. & Pradet, A. (1986) *Plant Mol. Biol.* **7**, 321-329.

Rowland, L.J. & Strommer, J. (1986) *Molecular and Cellular Biology* **6**, 3368-3372.

Rowland, L.J., Chen, Y.-C. & Chourey, P.S. (1989) *Molecular and General Genetics* **218**, 33-40.

Russell, D.A., Wong, D. M.-L. & Sachs, M.M. (1990) *Plant Physiology* **92**, 401-407.

Sachs, M.M., Freeling, M. & Okimoto, R. (1980) *Cell* **20**, 761-767.

Schwartz, D. & Endo, T. (1966) *Genetics* **53**, 709-715.

Tanksley, S.D. & Jones, R.A. (1981) *Biochemical Genetics* **19**, 397-409.

Tihanyi, K., Fontanell, A. & Thirion, J.-P. (1989) *Biochemical Genetics* **27**, 719-730.

Xie, Y. & Wu, R. (1989) *Plant Molecular Biology* **13**, 53-68.

Soybean Transformation to study Molecular Physiology

James E. Bond and Peter M. Gresshoff

Plant Molecular Genetics. The University of Tennessee, Knoxville, TN 37901-1071

Introduction

Gene transfer for the improvement of soybean

Soybean (*Glycine max*) is one of the world's most important agronomic crops. As such it has been the focus of extensive efforts towards genetic improvement both by conventional breeding techniques and genetic engineering. Conventional breeding has yielded impressive improvements in overall yield, oil composition, protein content, and resistance to disease. However, as in many other highly bred crop plants, modern soybean cultivars are based on a very limited gene pool. North American cultivars are derived from only 11 introductions (Allen and Bhardwaj, 1987). The vulnerability of crops to genetic uniformity was demonstrated in 1970 during the Southern corn leaf blight. In the near future the majority of genetic variation available for continued progress in soybean breeding (especially yield increase) will come from collections maintained in gene pools and cultivars in current production (Harlan, 1975; Hawkes, 1983). However, only limited genetic variation exists in areas such as disease and pest resistance as well as drought and salt tolerance.

Biotechnology promises other routes for genetic improvement in these areas.

0-8493-8263-7/93/$0.00 + $.50

Somaclonal variation from tissue culture of plants can be one useful source of genetic variation (Larkin et al, 1984). The culture procedure offers an efficient selection system for traits such as disease and herbicide resistance. A second route involves the use of tissue culture coupled with transformation of plants to introduce new genetic material. Transformation overcomes sexual barriers for the introduction of new genes from different plant genera or other kingdoms. The theoretical increase in genetic diversity of a crop plant is limitless. Transformation is not only useful in the addition of new genes, but also in the manipulation of endogenous genes, and characterization of the unknown genes. These methods will lead to better genetic, molecular and biochemical characterization of crop plants, enabling more precise manipulation and overall genetic improvement.

Progress in soybean transformation

Soybean has historically been a difficult crop to transform. Early attempts to transform soybean were confounded by difficulties in tissue culture as well as the limited susceptibility of cultivars to the classical plant transformation agent, *Agrobacterium tumefaciens*. Transformation via electroporation of naked DNA into protoplasts was also hindered by the inability to regenerate whole plants. Soybean transformation has only become possible after a long and difficult commitment from many research groups attempting to find suitable culture and gene transfer techniques.

This review will describe the early tissue culture work that was so important in developing successful gene transfer procedures, and the transformation methods and applications currently in use. Transformation of plants provides the possibility to monitor the expression of genes as they are affected by developmental and environmental factors.

Tissue culture of soybean

In the past few years various tissue culture systems have been developed for soybean. These include shoot morphogenesis from primary leaves, cotyledonary nodes and epicotyls (Barwale et al, 1986; Wright et al, 1986, 1987a,b), or somatic embryogenesis from immature zygotic embryos (cotyledons or embryo axis) (Christianson et al, 1983; Ranch et al, 1985). The cotyledonary node regeneration system was used by Hinchee et al (1988) to transform soybean plants using *Agrobacterium tumefaciens*. However, this system was hampered by low efficiency due to the infrequent transformation of regenerable cell types. The application of protoplast culture to the field of soybean transformation was slowed by the inability to regenerate whole plants. Lin et al (1987) and Christou et al (1988) demonstrated transformation of soybean protoplasts and the production of transgenic calli but were unable to regenerate plants. Dhir et al (1991) since reported plant regeneration from protoplasts of soybean, but the frequency of regeneration was low. It was not until 1992 that Dhir et al regenerated whole transgenic plants from protoplasts. This system has limitations due to the inefficiency of the regeneration system and its genotypic dependency.

Initiation of embryogenic suspension cultures (cv. 'Fayette') and subsequent regeneration of whole plants has been achieved from immature zygotic embryo axes (Christianson et al, 1983) and immature cotyledons (Finer and Nagasawa, 1988). These suspension cultures were initiated by culturing immature cotyledons to produce embryogenic cell clumps capable of producing small secondary embryoids. This tissue was used to initiate embryogenic suspension cultures termed 'highly embryogenic ontogenetic stage' ('early staged') suspensions. However, it has been reported that the first transgenic plants regenerated from these cultures were sterile. Regeneration systems such as the embryogenic suspension (Finer and Nagasawa, 1988) and protoplast regeneration systems (Dhir et al, 1991) are prone to produce regenerant plants showing somaclonal variation. Larkin et al (1984) saw somaclonal variation in tissue culture derived plants as early as 1984. This is thought to be a problem associated with tissue in long term culture. Freshly initiated suspensions of cv. Fayette gave fertile plants (pers. comm. Dr. J. Finer, Ohio State University).

Our study has initiated embryogenic suspension cultures of cv. Bragg and its derived supernodulating mutant nts1007 (see Carroll et al, 1985b). Whole fertile plants have been regenerated from these cultures. Bailey and Parrott (1992) also reported initiation of embryogenic suspension cultures and regeneration of fertile plants from cv. Century, Davis, Lee, Hutcheson, Peking and line PI 417138.

Embryogenic suspension cultures have proven to be efficient regeneration systems when coupled with particle gun transformation (Finer and McMullen, 1991). The reasons for the success with embryogenic suspension cultures is twofold. First, embryo development is initiated from single surface cells of older embryos which are readily accessible for particle gun bombardment. Second, the suspension conditions and the descent of subsequent embryos allows efficient selection of transgenic material via antibiotic selection. This is in contrast to the mixed results reported with bombardment of meristem tissue. McCabe (1988) reported regeneration of whole chimeric transgenic plants. Buising (1990) reported that their extensive efforts bombarding meristematic tissue did not yield transgenic plants. These mixed results highlight the difficulty of transforming the meristematic cells responsible for regeneration and germ line transfer.

Agrobacterium tumefaciens transformation

Agrobacterium tumefaciens has been used successfully to transform a wide range of dicotyledonous plants (Figure 1). This has been achieved by utilizing disarmed vectors to insert foreign genes (Bevin, 1984). However, susceptibility of different species and even cultivars to *Agrobacterium* varies greatly. Soybean is not very susceptible to *Agrobacterium* (DeCleene and DeLey, 1976).

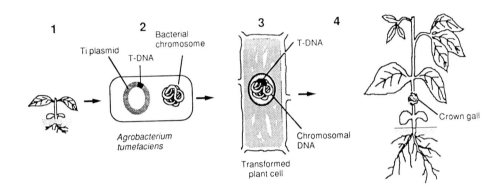

Figure 1: Transformation of plant tissue by the plant pathogen Agrobacterium tumefaciens. A virulent bacterium enters the plant at a wound site[1]. The Agrobacterium harbors a Ti plasmid in addition to the bacterial chromosome[2]. A region of the Ti plasmid called the T-DNA is transferred into the plant cell. This becomes integrated into the plant chromosome[3]. This transferred DNA encodes genes for the production of cytokinins and opines. The cytokinins induce cell divisions which eventually form a gall or tumor. In Agrobacterium rhizogenes auxin genes or genes which make the plant more susceptible to auxins are transferred. These genes stimulate the production of roots instead of galls. The opines serve as metabolites for the Agrobacterium living in the intercellular spaces of the gall.

The compatibility or virulence of a specific *Agrobacterium* is dependent on the genetic background of both host and pathogen. Thus much of the work in the early and mid-eighties concentrated on finding compatible *Agrobacterium*/plant associations. DeCleene and DeLey (1976), Pedersen et al (1983), Owens and Cress (1985), and Bryne et al (1987) showed that induction of tumors in soybean tissue was possible, but showed significant variation between cultivars.

Facciotti et al (1985) were able to introduce a light inducible chimeric gene into soybean tissue but could not regenerate plants. Byrne et al (1987) screened a large number of soybean genotypes and found cultivar Peking highly susceptible to the *Agrobacterium tumefaciens* nopaline strain A208. The soybean research group at Monsanto engineered a disarmed A208 strain which could be modified to harbor marker and selection genes either integrated into the Ti plasmid or located on a separate binary plasmid

(Hinchee et al, 1988). This vector was used to inoculate soybean petioles which were induced to produce multiple shoots as in the Barwale et al (1986) and Wright et al (1986a,b) culture procedures. It was soon found that the disarmed engineered A208 strain had much lower virulence than the non-disarmed strain making transformation *in vitro* and regeneration of whole transgenic plants difficult. An extensive tissue culture effort was required to overcome this problem This was compounded by the rarity of transformation and regeneration events occurring in the same cell types.

Eventually, successful transformation of whole plants was reported with the susceptible soybean cultivar Peking (Hinchee et al, 1988). Low transformation frequency (4% of plants regenerated through selection) and difficulties with reproducibility in other research groups coupled with the limited application to cultivars other than cv. Peking have suppressed the use of this system. Nevertheless, Townsend et al (1992) also reported successful transformation of soybean using this system.

Chee et al (1989) reported transformation of soybean cv. A0949 by infecting germinating seeds with *Agrobacterium tumefaciens* strain C58Z707. However, inheritance of the neomycin phosphotransferase (NPT II) gene was extremely infrequent (0.07% of infected seedlings showed inheritance to the R1 generation). Zhou and Atherly (1990) reported transformation of the maize controlling element (Ac) into soybean plants but inheritance to the progeny was not demonstrated. This work was repeated by Muhawish and Shoemaker (1992; including molecular evidence for inheritance).

Problems associated with Agrobacterium transformation systems

Agrobacterium naturally invades plant tissue and persists in the intercellular spaces and vascular tissues of the plant. This can complicate the detection of true transformation events when using this pathogen for gene transfer. Most commonly used marker and selection genes (GUS: ß-glucuronidase; CAT: Chloramphenicol acetyltransferase; NPT II: Neomycin phosphotransferase II, HYGRO: Hygromycin phosphotransferase) are positioned within the T-DNA borders and driven by eukaryotic promoters. Initially it was believed that the genes within these borders were not transcribed or translated in the *Agrobacterium*. Also it was assumed that the eukaryotic promoters did not function in the prokaryote. Both of these postulations were later found to be false (Vancanneyt et al, 1990). *Agrobacterium* has clearly been shown to express most commonly used marker and selection genes giving false positives in most tests and assays for transgenic tissue. With this in mind, any report of transformation should be reviewed critically.

The problems associated with assays that could not differentiate between bacterial contamination and true transformation events were overcome in two ways. First, many research groups turned towards particle gun gene transfer.

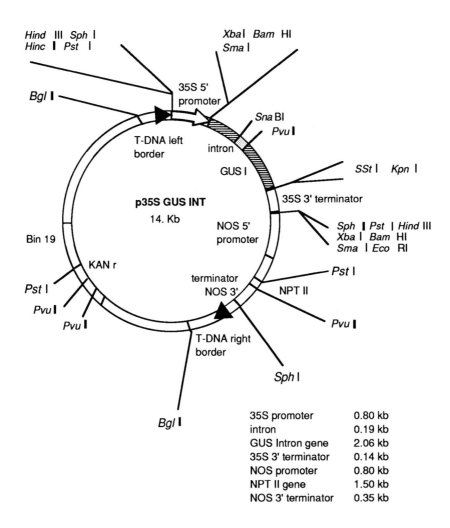

35S promoter	0.80 kb
intron	0.19 kb
GUS Intron gene	2.06 kb
35S 3' terminator	0.14 kb
NOS promoter	0.80 kb
NPT II gene	1.50 kb
NOS 3' terminator	0.35 kb

Figure 2: The p35S GUS intron plasmid, a disarmed binary vector for use in soybean transformation. NPTII + neomycin phospho-transferase gene; gus = ß-glucoronidase gene. Borders = T-DNA border from Agrobacterium tumefaciens. Restriction mapping was done by Ms. Natasha Taranenko (PMG, UT. Knoxville).

Second, other groups looked for ways to modify marker or selection genes to eliminate bacterial expression. One such gene that has been modified is the ß-glucuronidase gene (GUS). GUS has widely been used as a reporter gene in transformation systems (Jefferson 1987a,b), but it has also suffered from the ability of *Agrobacterium* to transcribe and translate the gene (Vancanneyt et al, 1990).

Vancanneyt et al (1990) introduced a plant intron in the coding region of the GUS gene (Figure 2). *Agrobacterium* lacks the RNA splicing apparatus required to remove this intron and produce an RNA transcript that can be translated. Thus the bacteria are not able to express GUS activity. Only when the T-DNA harboring the GUS gene is transferred to the eukaryotic cell will the GUS gene be expressed. This construct is now used extensively to test for transformation. To date there have been no published reports that show bacterial splicing of the gene and expression of GUS activity.

Southern hybridization of transgenic material

Southern hybridization is an important technique to demonstrate the integration of foreign DNA in transgenic plant tissue (Southern, 1975). However its use in *Agrobacterium tumefaciens* transformation must be viewed critically. Many reports show Southern blots where the plant genomic DNA is restricted with enzymes that cut twice inside the T-DNA borders of the inserted DNA. When probing the plant DNA with sequences internal to the T-DNA region a single or double band identical to one obtained from a plasmid digest will be obtained.

This shows that the sequence is present but does not indicate whether the introduced DNA is integrated, non-integrated, or is due to bacterial contamination. The correct procedure for Southern analysis is to probe the Southern blot with a border fragment. This means using a sequence that crosses a T-DNA border (preferably the right border as this border is essential for T-DNA transfer).

A restriction enzyme must be chosen which cuts once inside the T-DNA border and once outside the border. The genomic DNA should be cut with the same enzyme or enzymes. For example, in the p35S GUS I plasmid (Figure 2) this can be achieved by probing the blot with the *Bgl*II *Pvu*II fragment which contains part of the DNA from outside the T-DNA border as well as the 35S promoter and the GUS I gene. The genomic plant DNA should be digested with the same enzymes (*Bgl*II and *Pvu*II). If hybridization is seen only at 1.5 kb this shows either that there is bacterial contamination of the cultures or that the plasmid is present in the plant cell, but is not integrated. If hybridization is seen at different sized bands i.e. any larger or smaller than the expected 1.5 kb band, integration of the T-DNA into the plant genomic DNA would be proven. This is because the *Bgl*II site just outside the T-DNA should not be transferred but other *Bgl*II sites in the genomic plant DNA would be variable depending upon the transgene integration site in the plant DNA.

Because many plants including soybeans show low susceptibility to *Agrobacterium*, research has been undertaken on the induction of the virulence genes of *Agrobacterium* with specific media (Alt-Mörbe et al, 1989; Ankenbauer and Nester, 1990) as well as on the isolation, duplication and increase in expression of virulence genes (Jin et al, 1987).

Agrobacterium rhizogenes transformation

Agrobacterium rhizogenes is similar to *Agrobacterium tumefaciens*, but instead of being the causal agent of crown gall disease it elicits hairy root disease. This bacterium is virulent on a wide range of dicotyledonous plants (DeCleene and DeLey, 1981). It transforms plants by insertion of a region of DNA called T-DNA (transfer DNA) from the root-inducing (Ri) plasmid into the plant genome (Chilton et al, 1982; for review of the molecular biology of the Ri plasmid, see Sinkar et al, 1987).

The *A. rhizogenes* strains are classified by the opines which are synthesized by genes in the Ri T-DNA. Opines, which are conjugated amino acids and carboxylic acids are synthesized by these genes and serve as metabolites for the *Agrobacterium* (Petit et al, 1983). To date four different opine types have been found in *A. rhizogenes* induced roots. These are agropine, mannopine, cucumopine and mikimopine (Davioud et al, 1988; Filetici et al, 1987; Isogai et al, 1988; Petit et al, 1985).

In strains of *A. rhizogenes* that produce agropine the T-DNA is split into two regions, T_L and T_R, each approximately 15-20 kb in size. Both T-DNA regions are integrated separately into the plant genome (Phelep et al, 1991). Agropine strains have both the agropine synthesis and auxin indole acetic acid (IAA) synthesis genes referred to as *aux1* and *aux2* or *tms1* and *tms2* on the T_R segment. Cucumopine and mannopine *A. rhizogenes* strains have only one T-DNA region between 34 to 42 kb in size.

The function of rol genes in Agrobacterium rhizogenes transformation

The T_L segment of the agropine strain contains 18 open reading frames consisting of 255 nucleotides or more (Slightom et al, 1986). Transposon mutagenesis of this region has revealed the presence of at least four genetic loci (*rolA, rolB, rolC* and *rolD*) which also affect root induction. The *rol* (rooty loci) and *aux* genes are responsible for root induction either individually or together, depending on the plant species or type of tissue (White et al, 1985).

Interestingly cucumopine and mannopine *A. rhizogenes* strains do not have *aux* genes, only *rol* genes, and yet they still produce a hairy root phenotype. Also, although *aux* genes have been identified in agropine *A. rhizogenes* strains, it has been shown that their presence is not required for hairy root induction in tobacco (Vilaine et al, 1987). In fact it has also been shown that the entire T$_R$ fragment is not required for hairy root production. Transgenic lines elicited by the agropine-type *A. rhizogenes* strains do not always synthesize agropine (Petit et al, 1983). In these cases it seems that sufficient auxin is supplied by the plant and that the *aux* genes may play a role in supplementing auxin only when it exists in limiting amounts (Grierson and Covey, 1988).

The *rolA, rolB, rolC* and *rolD* loci affect virulence and hairy root formation. Originally it was thought that the *rol* genes synthesized auxins causing the characteristic hairy roots similar to the way that *Agrobacterium tumefaciens* initiate callus production by transforming cytokinin synthesizing genes (Zambryski et al, 1989). It now appears that the mechanism is much more complex. Shen et al (1988) reported that transformed roots of this type contained less auxin but were 10 to 100 times more sensitive to the presence of auxins than normal roots. Because these roots did not show altered, auxin-independent proton excretion, altered ethylene sensitivity or increased auxin uptake, they suggested that the transformed genes affected the auxin receptor-transduction system. Shen et al (1988) proposed that increased auxin sensitivity induced root induction. This result is supported by the work of Maurel et al (1991) with tobacco protoplasts. They reported that sensitivity to auxin by transformation of single *rol* genes could be increased 30 fold by the addition of all the *rol* genes on the T$_L$-DNA segment of an agropine strain.

Roots are known to be initiated by endogenous concentrations of auxins. However, Croes et al (1989) detected no differential auxin sensitivity between normal and transformed root cultures. Quattrochio et al (1986) showed that anti-auxins had an inhibitory effect on the growth of transformed potato roots. Cardarelli (1987) and Spena et al (1987) have demonstrated that when *rolA, rolB* and *rolC* genes were expressed individually, transgenic roots were still produced but at an attenuated level. The effects of each *rol* gene on root initiation and growth are different (Schmülling et al, 1988, 1989). The *rolB* and *rolC* genes seem to be involved in root branching, but the level and region of their expression are different. Expression of *rolB* is in the root cap whereas *rolC* is expressed in the phloem area (Schmülling et al, 1988). Mutations in the *rolA* gene (*rolA⁻*) result in the formation of long straight roots. A mutation in *rolB* eliminates both callus and root formation at the wound site. *The rolC* and *rolD* mutations are more subtle. Root growth is retarded with a *rolC* mutation, whereas a *rolD* mutation gives rise to *Agrobacterium tumefaciens*-like tumors (White et al, 1985). The primary determinate in the severely wrinkled leaf phenotype seen in *A. rhizogenes* transformed plants is the *rolA* gene (Sinkar et al, 1987).

Estruch et al (1991b) studied the RolC peptide and found that it was a non-transported 20kDa cytosolic protein. Plants expressing this protein showed dwarfing and reduced apical dominance. Chimeric plants for *rolC* expression showed leaves with pale green sectors with sharp borders, indicating lack of diffusion or transport of this protein. This is compatible with the hypothesis that plant growth and morphogenesis is altered by the *rol* gene proteins which modify endogenous plant hormones (Estruch et al, 1991a).

Estruch et al (1991b) found that the rolC protein releases cytokinins from glucoside conjugates. RolB protein was found to hydrolyze indole glucosides, thus releasing auxins from its conjugate (Estruch et al, 1991c). The combined action of these two genes is thought to modulate the intracellular concentration of auxins and cytokinins to produce 'rhizogenes' roots. Plant developmental and physiological processes including nodulation can be influenced by enzymatic systems leading to conjugation and deconjugation of auxin and cytokinin phytohormones.

Production of transgenic plants from Agrobacterium rhizogenes transformed roots

A. rhizogenes induced hairy roots of approximately 50 different plant species have been cultured. Transgenic plants have been regenerated from these roots either by spontaneous shoot formation or through a callus stage initiated on a medium supplemented with auxins and cytokinins (Potter, 1991).

Desired sequences can be transferred into plants using *A. rhizogenes* mediated transfer. This can be achieved either by inserting the DNA sequence into an *A. tumefaciens* derived binary plasmid (Simpson et al, 1986) or by integrating the sequence into the pRi T_L-DNA segment of the Ri plasmid (Stougaard et al, 1987a).

Phelep et al (1991) used the *A. rhizogenes* strain K599 as well as the agropine strain A4 to transform the woody non-legume *Allocasuarina*. Their study was characterized by careful interpretation of Southern hybridization data which showed covalent integration of *Agrobacterium* sequences into plant DNA. Limited success, however, has been reported in the *Glycine* genus. Only in the wild soybean *Glycine canescens* have whole plants been regenerated from transgenic roots (Rech et al, 1988).

Susceptibility of soybean to Agrobacterium rhizogenes

To date there are no reports of whole plants regenerated from *A. rhizogenes*-induced roots of soybean. Even production of these roots has proven difficult. Owens and Cress (1985) tested the response of 26 *G. max* genotypes to *A. rhizogenes* strain A136 harboring the pRiA4b (Ri) plasmid. Seven out of 26 of these genotypes produced roots at the infection site. Attempts to culture these roots failed and assays for 'hairy root' markers (opine content) were not performed. Simpson et al (1986) used *A. rhizogenes* roots to introduce transgenic sequences from *A. tumefaciens* via a binary plasmid conferring kanamycin resistance. This plasmid functioned in *A. rhizogenes* (agropine strain A4) and produced roots which had the transgenic sequence integrated and expressed in tomato, tobacco and alfalfa. Although Simpson et al (1986) demonstrated a relatively high frequency of transgenic roots in alfalfa (33% nopaline positive, 4% kanamycin resistant), tomato (33% nopaline positive, 19% kanamycin resistant) and tobacco (19% kanamycin resistant, nopaline not determined), in soybean only 2% of the roots were nopaline positive and none were kanamycin resistant. In *G. max* a high percentage of non-transgenic roots were produced from the inoculation/wound site (Simpson et al, 1986). The ability of soybean to show intrinsic nopaline activity (Christou et al, 1986), coupled with the fact that none of the roots showed kanamycin resistance seriously limited the use of *A. rhizogenes* to produce transgenic roots in soybean.

It was not until 1990 that Savka et al. found an *A. rhizogenes* strain (cucumopine strain K599/pRi2659) capable of inducing 'rhizogenes' roots at a high frequency in *G. max* (37% of the cotyledons produced roots on the 10 genotypes tested). Root cultures derived from these infection events were positive for cucumopine production and grew rapidly in culture. Savka et al (1990) used these cultures to propagate soybean cyst nematodes *in vitro* but did not evaluate their nodulation phenotype.

Application of gene transfer techniques to the genetics of nodulation

The isolation of several nodulation cDNA clones and the comparison of sequence homology and amino acid composition with other genes has defined the biochemical function of only a few nodulins (Nap and Bisseling, 1990). Many, like Enod2, elude precise functional assignment, although similarity to other proteins exist. Nodulin Enod2 was found to be similar to hydroxyproline-rich cell wall proteins and is proposed to function in the oxygen diffusion barrier of the nodule. Plant transformation will help to elucidate the role of these symbiotic plant genes (Delauney et al, 1988).

Stougaard et al (1987b) were the first researchers to use transgenic plants to study the genetics of nodulation. They demonstrated nodule-specific expression of a complete soybean leghemoglobin gene in the root nodules of transgenic *Lotus corniculatus*. Later work focused on using the 5' promoter sequence from the leghemoglobin gene fused to the CAT reporter gene in order to show tissue-specific expression in transgenic nodules of *Lotus corniculatus* (deBruijn et al, 1989; 1990; Jensen et al, 1988; Szabados et al, 1990). Similar work was carried out using the *Parasponia* hemoglobin promoter and GUS reporter gene fusions in transgenic *Lotus* and tobacco (Bogusz et al, 1990).

Verma et al (unpubl. data) used the same promoter-reporter gene fusion technique to study the tissue-specific expression of nodulin-26. They discovered nodule-specific expression in soybean, but unexpectedly expression in the region of emerging laterals in a heterologous host (Vigna). They suggested that soybean (the homologous host) contained a suppresser of N-26 expression in tissues other than the nodule. This suppresser putatively is absent in the heterologous host. Nodulin-26 is a major protein of the peribacteroid membrane. GUS expression was seen at emerging nodules and root primordia. In *Vigna aconitifolia*, roots expressing anti-sense N-26 showed retarded hairy root growth (*A. rhizogenes*). Anti-sense studies were also conducted with nodulin-35 (N-35). Soybean nodulin-35 encodes a subunit of uricase localized in the peroxisomes of uninfected nodule cells (Nguyen et al, 1985). Lee et al (1992) introduced an anti-sense cDNA clone into *Vigna* behind a CaMV-35S promoter. Nodules formed on the resultant transgenic roots were smaller in size and plants exhibited a yellow leaf phenotype. Uricase activity was reduced by 50-60% as shown by enzymatic assays. Electron microscopy also showed that the size of the peroxisomes was reduced.

To date few transformation systems have been used for the study of symbiotic nitrogen fixation in soybean. In part this is because no suitable transformation system exists for such a study in soybean. Research in our laboratory has demonstrated that transgenic roots of several soybean lines can be produced with relative ease. We showed that the transgenic roots of soybean nodulated with normal nodule distribution and numbers, being 'true' to the genotype of the original seed line (e.g., supernodulation, non-nodulation).

The reporter gene was combined with the 35S promoter from cauliflower mosaic virus and is terminated by the nopaline synthase 3' region. Expression was noted in all tissue types, other than the bacteria-infected central zone of the nodule. Low GUS expression may be caused by the difficulty of oxygen diffusion needed for blue color development.

These nodules fix nitrogen and the degree of nodulation is still conditioned by the host's genotype (Table 1). This is contrary to the results of Beach and

Gresshoff (1986) where wild-type transgenic roots of siratro and red clover induced, by *A. rhizogenes*, showed nodulation interference or inhibition. There is no explanation for this discrepancy at the moment, although the nodule ontogeny, with the related differences in physiology as well as *Agrobacterium* strain differences may account for some variation.

Table 1. The mean nodule number (± SE) in transgenic (A. rhizogenes induced roots) and non-transgenic root systems (wild-type roots) of nodulation mutants nts382, nod139, nod49 and wild-type plants Bragg and Peking. Five plants were used per sample and the nodule number was measured 4 weeks after inoculation with Bradyrhizobium japonicum USDA 110.

Plant type	Nodule number per plant	
	non-transgenic root system	transgenic root system
nts382	255 ± 75	246 ± 39
nod49	0	0
nod139	0	0
Bragg (wild-type)	26 ± 6	24 ± 7
Peking (wild-type)	12 ± 3	12 ± 2

Not all nodulated root systems showed nodules with normal determinate soybean nodule morphology. About 15 to 20% of the transgenic root systems developed nodules, similar to those seen on pea or alfalfa. Such nodule phenotypes was never observed in this laboratory. The nodules were not puffy or callous-like as sometimes observed with incompatible infections. The nodules maintained a proper external layer, characterized by lenticels. The interior contained normal soybean nodule features such as scleroidal cells, nodule cortex and parenchyma, infected and uninfected cells. The essential difference was the presence of multiple active meristems in the nodule cortex as detected by microscopy. Infection and internal nodule development appeared normal, but early senescence sat in, leading to an early cessation of nitrogen fixation and loss of leghemoglobin. These nodules are elongated and often appear branched. The exact mechanism by which the phenotype is altered is not known at present, although the involvement of the *rol* genes from strain K599 and their effect on phytohormone levels is suspected (Bond et al, 1993; Gresshoff, 1993).

Research in our laboratory has focused on the plant genes involved in the nodulation process. The production of a series of soybean nodulation mutants (supernodulating nts382 and non-nodulating nod49 and nod139 lines) via EMS mutagenesis (Carroll and Gresshoff, 1985a,b; 1986; Gresshoff and Delves, 1986) has led to research aimed at finding the genes responsible for this changed phenotype. Any candidate sequences found will have to be tested via transformation to prove functionality to these genes. This will be performed as in classical complementation experiments, The candidate wild-type sequence will be transformed into the mutant plant to determine if the nodulation phenotype is corrected.

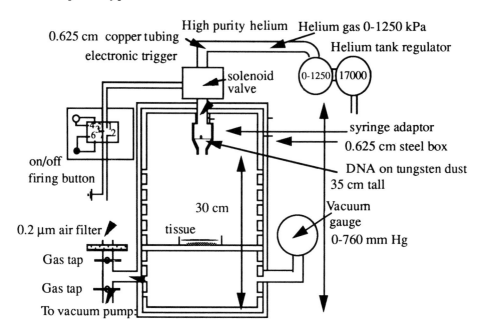

Figure 3: The UT-built helium micro-projectile gun. Before the bombardment of plant tissue, a 2 μl DNA-coated tungsten particles was suspended on a screen in the stainless steel syringe adapter. Helium was supplied through a tank regulator in the range of 0 to 1250 kPa He pressure. A short 50 milli-second burst of helium (normally 500 kPa) enters the chamber through an electronically triggered solenoid. This propelled the tungsten particles creating a fine aerosol. DNA carried into the nucleus was released from the tungsten particles to permit transient gene expression or integration into the plant genome and stable gene expression (Yamashita et al, 1991). Figure modified from Bond et al, 1992).

We are also interested in cloned and characterized plant genes such as chalcone synthase and nodulin-26 and their possible involvement in nodulation. Genes which are thought to be important in the nodulation process are being tested by antisense technology to determine if these genes are indeed involved in this process.

Micro-projectile 'gene' gun technology

In recent years, plant transformation strategies have come to rely on micro-projectile gun delivery of DNA to plant cells (see Figure 3). This process also called the 'biolistic' method was developed to overcome the transformation barriers of species and cell types. The technology uses micron-sized DNA-coated tungsten (or gold) particles accelerated to high velocities for penetration into plant cells and DNA delivery. This approach was applied to obtain and examine gene expression of both transient (i.e. short-term) and stable (i.e. inheritable) transformations of plant cells. In some cases whole transformed plants have been regenerated (see Mendel, 1990 for a review). More commonly, transformations have involved tissue-cultured cells transiently expressing reporter genes such as those encoding chloramphenicol acetyltransferase (CAT; Klein et al, 1988*a*) or ß-glucuronidase (GUS; Jefferson, 1987a,b).

Christou et al (1988) were the first to transform soybean tissue by particle bombardment of immature embryos, which gave rise to kanamycin-resistant protoplasts and subsequent callus. This breakthrough was followed by stable transformation and regeneration of soybean plants from immature embryo axes bombarded with a GUS construct (McCabe et al, 1988; Christou et al, 1989). More recently, Finer and McMullen (1991) transformed soybean by particle bombardment of embryogenic suspension cultures. Similar results were obtained in our laboratory both with cv. Fayette (as used by Finer and McMullen, 1991), but also with cv. Bragg and its supernodulation mutant. Other labs have been able to reproduce this result. For example, Parrott et al (1992) introduced the *Bacillus thuringiensis* (BT) toxin gene into soybean using embryogenic suspension cultures and particle bombardment.

Future applications of soybean transformation

Soybean transformation is being used commercially to introduce genes conferring herbicide tolerance, improved protein quality and composition as well as disease and pest resistance. The Monsanto Company has already produced soybean (cv. Peking) plants resistant to the herbicide glyphosate (Round Up™). The BT gene has been introduced into soybean and conferred increased resistance to velvet bean caterpillar, a common soybean pest (Parrot et al, 1992). Also, the methionine-rich seed storage protein (Bex) from brazilnut has been transformed into soybean to improve the seed protein quality (Townsend et al, 1992). Soybean mosaic virus resistance may be conferred to commercial cultivars by the transformation and expression of the viral coat protein in the plant. Similar commercial applications in superior soybean cultivars will be seen in the future.

From the academic point of view there are many fundamental questions which can be answered using gene transfer techniques (transient and stable transformation systems). Modifications of plants by gene addition or manipulation of endogenous genes will help elucidate the action of specific sequences. The analysis of promoter sequences, using reporter gene fusions, will help elucidate the timing and tissue specificity of gene expression. Understanding what these sequences do and when and where they express will lead to better genetic, molecular and biochemical characterization of soybean. This will enable precise manipulation and genetic improvement of this valuable crop plant. Some of the systems that will be used for this include site-directed mutagenesis, transposon tagging of genes, and ribozyme and antisense technology. Transformation could be valuable to find the genes mutated in plant mutants by classical complementation.

Acknowledgments

This study was supported by the fund for the Ivan Racheff Chair of Excellence in Plant Molecular Genetics, the Human Frontiers Science Programme, the Tennessee Soybean Promotion Board, and the United Soybean Board (ASA).

References

Allen, F.L. & Bhardwaj, H.L. (1987) Genetic relationships and selected pedigree diagrams of North American soybean cultivars. *University of Tennessee Agricultural Experimental Station Bulletin* 652.

Alt-Mörbe, J., Kühlmann, H. & Schröder, J. (1989) *Mol. Plant-Microbe Interactions* 2, 301-308.

Ankenbauer, R.G. & Nester, E.W. (1990) *J. Bacteriol.* 172, 6442-6446.

Bailey, M.A. & Parrott, W.A. (1992) *Proceedings of 4th Biennial Conference on Molecular and Cellular Biology of the Soybean.* Iowa State University, Ames Iowa.

Barwale U.B., Kerns H.R. & Widholm J.M. (1986) *Planta* 167, 473-481.

Bevin, M. (1984) *Nucl. Acids Res.* 12, 8711-8721.

Bond, J.E., McDonnell, R.E., Finer, J.J. & Gresshoff, P.M. (1992) *Tenn. Farm & Home Sci.* 162, 4-14.

Bond, J.E., McDonnell, Farrand, S., Joshi, P., & Gresshoff, P.M. (1993) Meristematic nodule morphology in soybean after *Agrobacterium rhizogenes* transformation of roots (in preparation).

Buising, C.M. (1990) *Proceedings of 3rd Biennial Conference on Molecular and Cellular Biology of the Soybean.* Iowa State University, Ames Iowa.

Byrne, M.C., McDonnell, R.E., Wright, M.S. & Carnes, M.G. (1987) *Plant Cell Tissue and Organ Culture.* 8, 3-15.

Cardarelli, M., Mariotti, D., Pomponi, M., Spano, L., Capone, I. & Constantino, P. (1987) *Mol. Gen. Genet.* 209, 475-480.

Carroll, B.J., McNeil, D.L. & Gresshoff, P.M. (1985a) *Proc. Natl. Acad. Sci. (USA)* 82, 4164-66.

Carroll, B.J., McNeil, D.L. & Gresshoff, P.M. (1985b) *Plant Physiol.* 78, 34-40.

Carroll, B.J., McNeil, D.L. & Gresshoff, P.M. (1986) *Plant Science* 47, 109-114.

Chee, P.P., Fober, K.A. & Slightom, J.L. (1989) *Plant Physiol.* 91, 1212-1218.

Chilton, M.D. (1983) *Sci. American* **248**, 50-59.

Chilton, M.D., Tepfer, D.A., Petit, A., David, C., Casse-Delbart, F. & Tempé, J. (1982) *Nature* **295**, 432-434.

Christianson M L, Warnick, D.A & Carlson, P.S. (1983) *Science* **222**, 632-634.

Christou, P., Platt, S.G. & Ackerman, M.C. (1986) *Plant Physiol.* **82**, 218-221.

Christou, P., McCabe, D.E. & Swain, W.F. (1988) *Plant Physiol.* **87**, 671-674.

Christou, P., Swain, W.F., Yang, N.S. & McCabe, D.E. (1989) *Proc. Natl. Acad. Sci. (USA)* **86**, 7500-7504.

Croes, A.F., van den Berg, A.J.R., Bosveld, M., Breteler, H. & Wullems, G.J. (1989) *Planta* **179**, 43-50.

Davioud, E., Quirion, J-C., Tate, M.E., Tempé, J. & Husson, H-P. (1988) *Heterocycles* **27**, 2423-2430.

deBruijn, F.J., Felix, G., Grunenberg, B., Hoffman, H.J., Metz, B., Ratet, P., Simons-Schreier, A., Szabados, L., Welters, P., & Schell, J. (1989) *Plant Mol. Biol.* **13**, 319-325.

deBruijn, F.J., Szabados, L. & Schell, J. (1990) *Dev. Genet.* **11**, 182-196.

DeCleene, M. & DeLey, M. (1976) *Bot. Rev.* **42**, 389-486.

DeCleene, M. & DeLey, M. (1981) *Bot. Rev.* **47**, 147-194.

Delauney, A.J., Tabaeizadeh, Z. & Verma, D.P.S. (1988) *Proc. Natl. Acad. Sci. (USA)* **85**, 4300-4304.

Dhir, S.K., Dhir, S. & Widholm, J.M. (1991) *Plant Cell Reports* **10**, 39-43.

Dhir, S.K., Dhir, S., Savka, M.A., Belanger, F., Kriz, A.L., Farrand, S.K. & Widholm, J.M. (1992) *Plant Physiol.* **99**, 81-88.

Estruch, J.J., Parets-Soler, A., Schmülling, T. & Spena, A. (1991a).*Plant Mol. Biol. Rep.* **17**, 547-550.

Estruch, J.J., Chriqui, D., Grossman, K., Schell, J. & Spena, A. (1991b) *EMBO. J.* **10**, 2889-2895.

Estruch, J.J., Schell, J. & Spena, A. (1991c) *EMBO. J.* **10**, 3125-3128.

Facciotti, D., O'Neal, J.K., Lee, S. & Shewmaker, C.K. (1985) *Bio/Technology.* **3**, 241-246.

Filetici, P., Spano, L. & Costantino, P. (1987) *Plant Mol. Biol.* **9**, 19-26.

Finer, J.J. & Nagasawa, A. (1988) *Plant Cell Tissue and Organ Culture* **15**, 125-136.

Finer, J.J. & McMullen, M.D. (1991) *In Vitro Cell & Devel. Biol.* **27**, 175-182.

Gresshoff, P.M. (1993) In: *Nitrogen Fixation.* eds. R. Palacios, J. Mora and W.E. Newton. Kluwer

Publ. Comp. Dortrecht, The Netherlands, (in press).

Grierson, D. & Covey, S.N. (1988) *Plant Molecular Biology.* 2nd edition. Chapter 7, Chapman and Hall, New York.

Harlan, J.R. (1975) *Crops and Man.* American Society of Agronomy, Crop Science Society of America, Madison, Wisconsin.

Hawkes, J.G. (1983). *The Diversity of Crop Plants.* Harvard University Press, Cambridge, Massachusetts.

Hinchee, M.A.W., Conner-Ward, D.V., Newell, C.A., McDonnell, R.E., Sato, S.J., Gasser, C.S., Fischoff, D.A., Re, D.B., Fraley, R.T. & Horsch, R.B. (1988) *Bio/Technology* **6,** 915-922.

Isogai, A., Fukuuchi, N., Hayashi, M., Kamada, H., Harada, H. & Suzuki, A. (1988) *Agric. Biol. Chem.* **52,** 3235-3238.

Jefferson, R.A. (1987a) *Plant Mol. Biol. Rep.* **5,** 387-405.

Jefferson, R.A. (1987b) *EMBO. J.* **6,** 3901-3907.

Jensen, E.O., Marcker, K.A., Schell, J. & deBruijn, F.J. (1988) *EMBO. J.* **7,** 1265-1271.

Jin, S., Komari, T., Gordon, M.P. & Nester, E.W. (1987) *J. Bacteriol.* **169,** 4417-4425.

Klein, T.M., Wolf, E.D., Wu, R. & Sanford, J.C. (1987) *Nature* **327,** 70-73.

Klein, T.M., Fromm, M., Weissinger, A., Tomes, D., Schaaf S., Sletten, M. & Sanford, J. C. (1988a) *Proc. Natl. Acad. Sci. (USA).* **85,** 4305-4309.

Landau-Ellis, D., Angermüller, S., Shoemaker, R. & Gresshoff, P.M. (1991) *Mol. Gen. Genet.* **228,** 221-226.

Larkin, P.J., Ryan, S.A., Brettell, R.I.S. & Scowcroft, W.R. (1984) *Theor. Appl. Genet.* **67,** 443-455.

Lee, N.G., Stein B. & Verma, D.P.S. (1993) *Plant Journal* (in press).

Lin, W., Odell, J. T. & Schreiner, R.M. (1987) *Plant Physiol.* **84,** 856-861.

Maurel, C., Barbier-Byrgoo, H., Spena, A., Tempé, J. & Guern, J. (1991) *Plant Physiol.* **97,** 212-216.

McCabe, D.E., Swain, W.F., Martinell, B.J. & Christou, P. (1988) *Bio/Technology* **6,** 923-926.

Mendel, R.R. (1990) *AgBiotech News and Information* **2,** 643-645.

Miao, G-H., Hirel, B., Marsolier, M.C., Ridge, R.W. & Verma, D.P.S. (1991) *Plant Cell* **3,** 11-22.

Muhawish, S.M. & Shoemaker, R.C. (1992) *Proceedings of the 4th Biennial Conference on Molecular and Cellular Biology of the Soybean.* Iowa State University, Ames Iowa.

Owens, L.D. & Cress, D.E. (1985) *Plant Physiol.* **77,** 87-94.

Parrott, W.A., Bailey, M.A., Adang, M.J., Boerma, H.R. & All, J.N. (1992) *Proceedings of the 4th Biennial Conference on Molecular and Cellular Biology of the Soybean.* Iowa State University, Ames Iowa.

Pedersen, H.C., Christianson, J. & Wyndaele, R. (1983) *Plant Cell Reports* 2, 201-204.

Petit, A., David, C., Dahl, G., Ellis, J.G., Guyon, P., Casse-Delbart, F.C. & Tempé, J. (1983) *Mol. Gen. Genet.* 19, 204-214.

Petit, A. & Tempé, J. (1985) In: *Molecular Form and Function of the Plant Genome.* (Van Vloten-Doting, L., Groot, G.S.P., & Hall, T.G., eds) pp 625-636, Plenum Publishing Co, New York.

Petit, A., Stougaard, J., Kuhle, A., Marcker, K.A. & Tempé, J. (1987) *Mol. Gen. Genet.* 207, 245-250.

Phelep, M., Petit A., Martin, L., Duhoux, E. & Tempé, J. (1991) *Bio/Technology* 9, 461-466.

Potter, J.R. (1991) *Critical Reviews in Plant Science*s 10, 387-421.

Quattrochio, F., Benvenuto, E., Tavazza, R., Cuozzo, L. & Ancora, G. (1986) *J. Plant Physiol.* 123, 143-150.

Ranch, J.P., Oglesby, L. & Zielinski, A.C. (1985) *In Vitro Cell & Dev. Biol.* 21, 653-658.

Rech, E.L., Golds, T.J., Hammatt, N., Mulligan, B.J. & Davey, M. (1988) *J. Exp. Bot.* 39, 1275-1285.

Rolfe, B.G. & Gresshoff, P.M. (1988) *Annu. Rev. Plant Physiol. Plant Mol. Biol.* 39, 297-319.

Savka, M.A., Ravillion, B. Noel, G.R. & Farrand, S.K. (1990) *Phytopathology* 80, 503-508.

Schmülling, T., Schell, J. & Spena, A. (1988) *EMBO. J.* 7, 2621-2629.

Schmülling, T., Schell, J. & Spena, A. (1989) *Plant Cell* 1, 665-670.

Shen, W.H., Petit, A., Guern, J. & Tempé, J. (1988) *Proc. Natl. Acad. Sci. (U.S.A.)* 85, 3417-3422.

Simpson, R.B., Spielmann, A., Margossian, L. & McKnight, T.D. (1986) *Plant Mol. Biol.* 6, 403-415.

Sinkar, V.P., White, F.F. & Gordon, M.P. (1987) *J. Bioscience* 11, 47-57.

Slightom, J.L., Durand-Tardif, M., Jouanin, L. & Tepfer, D. (1986) *J. Biol. Chem.* 261:108-121.

Southern, E.M. (1975) *J. Mol. Biol.* 98, 503-517.

Spena, A., Schmülling, T., Koncz., C. & Schell, J. (1987) *EMBO. J.* 6, 3891-3899.

Stougaard, J., Abildsten, D. & Marcker, K.A. (1987a) *Mol. Gen. Genet.* 207, 251-255.

Stougaard, J., Petersen T E. & Marcker, K.A. (1987b)*Proc. Natl. Acad. Sci. (USA)* 84, 5754-5757.

Townsend, J.A., Thomas, L.A., Kulisek, E.S., Daywalt, M.J., Winmter, K.R.K. & Altenbach,

S.B. (1992) *Proceedings of 4th Biennial Conference on Molecular and Cellular Biology of the Soybean.* Iowa State University, Ames Iowa.

Vancanneyt, G., Schmidt, R., O'Connor-Sanchez, A., Willmitzer, L. & Rocha-Sosa, M. (1990) *Mol. Gen. Genet.* **220**, 245-250

Vilaine, F. & Casse-Delbart, F. (1987) *Mol. Gen. Genet.* **206**, 17-23.

White, F.F., Taylor, B.H., Huffmann, G.A., Gordon, M.P. & Nester, E. W. (1985) *J. Bacteriol.* **164**, 33-44.

Wright, M.S., Koehler, S.M., Hinchee, M.A. & Carnes M.G. (1986) *Plant Cell Reports* **5**, 150-154.

Wright M.S,. Ward, D.V., Hinchee, M.A., Carnes, M.G. & Kaufman, R.J. (1987a) *Plant Cell Reports*
6, 83-89.

Wright, M S., Williams, M.H., Pierson, P.E. & Carnes, M.G. (1987b) *Plant Cell Tissue and Organ Culture* **8**, 83-90.

Yamashita, T., Iida, A. & Morikawa, H. (1991) *Plant Physiol.* **97**, 829-831.

Zambryski, P., Tempé J. & Schell, J. (1989) *Cell* **56**, 193-200.

Zhou, J.H. & Atherly, A.G. (1990) *Plant Cell Reports* **8**, 542-545.

Rhizobium Lipo-oligosaccharides: Novel Plant Growth Regulators

Gary Stacey[1], Juan Sanjuan[1], Herman Spaink[2], Ton van Brussel[2], Ben J.J. Lugtenberg[2], John Glushka[3] and Russell W. Carlson[3]

[1]*Center for Legume Research, Department of Microbiology, The University of Tennessee, Knoxville, TN 37996-0845, USA*
[2]*Leiden University, Institute of Molecular Plant Sciences, Nonnensteeg 3, 2311 VJ Leiden, The Netherlands*
[3]*Complex Carbohydrate Research Center, The University of Georgia, Athens, GA 30602, USA*

Introduction

The establishment of a nitrogen-fixing symbiosis between rhizobia (i.e., *Rhizobium, Azorhizobium* or *Bradyrhizobium* species) and legumes is a complex, multi-step process. Initially, rhizobia in the rhizosphere are chemoattracted to compounds excreted by legume roots. Flavonoids excreted by the plant are recognized by the bacteria and act to induce transcription of nodulation genes (*nod/nol*) that are required for plant invasion. The bacteria attach to the root hair and induce curling of the hair. The bacteria penetrate the hair cell and induce the formation of an infection thread (IT). Before and during IT formation, *Rhizobium* factors (i.e., lipo-oligosaccharides) act as signals to initiate root hair curling and nodule meristem formation. The IT grows in the direction of the meristem and ramifies into the plant cortex until it reaches a host cell where the bacteria are released. The bacteria released into the plant cell are contained in symbiosomes, the functional unit for nitrogen fixation (Roth et al, 1988). Subsequent to release, numerous metabolic processes are integrated to produce needed substances to maintain the symbiosis. The plant host undergoes several developmental changes leading to the formation of the nodule structure. Indeed, during the infection process, a number of gene products, termed nodulins, are specifically expressed (reviewed in Govers et al, 1987; Verma and Long, 1983; Verma et al, 1986; Gloudemans and Bisseling, 1989). Likewise, in addition to the *nod* genes which are required for nodule formation, the bacteria specifically express *nif* and *fix* genes which are necessary for nitrogen fixation or maintenance of the symbiotic state. Finally, depending on the plant species, the nodule can senesce, i.e., all of the symbiosomes and host cells are degraded. At this stage, it is likely that bacteria remaining in the infection thread utilize the decaying plant material as a food source until they are released into the soil environment.

This short description of the life and death of a nodule does not do justice to the beauty and complexity of this developmental process. Besides its practical importance due to the agronomic benefit of symbiotic nitrogen fixation, nodule development is a marvelous system for the study of basic questions of bacterial and plant development. One area that has received recent research emphasis is the definition of communication pathways between the bacterial symbiont and plant host. This communication serves to induce important developmental events and to coordinate bacterial-plant interaction.

Role of the *nod* genes in legume infection

Long et al (1982) and Scott et al (1982) were the first to isolate *Rhizobium* genes essential for nodule formation. Since that time, progress in identification of genes essential for nodulation has been rapid. The *nodDABCIJ* genes are "common" *nod* genes that show significant homology among *Rhizobium, Azorhizobium* and *Bradyrhizobium* species. The *nodD* gene encodes a positive regulatory protein required for expression of other *nod* genes (for example, Horvath et al, 1987; Mulligan and Long, 1985; Rossen et al, 1985; Shearman et al, 1986). NodD activates transcription only in the presence of host produced flavonoids. In addition to the common *nod* genes, each *Rhizobium* possesses other genes which seem to be important in the ability to selectively infect certain legume species. Examples of these genes are the *nodEFGHQ* genes of *R. meliloti* (Kondorosi et al, 1985; Kondorosi et al, 1984), *nodEFLMN* genes of *R. leguminosarum* (Downie et al, 1983; Hombrecher et al, 1984; Djordjevic et al, 1985; Schofield et al, 1984) and the *nodZ* gene of *B. japonicum* (Stacey et al, in preparation).

Evidence is now accumulating to suggest that many of the *nod* genes, if not all, are involved directly or indirectly in the production of novel lipo-oligosaccharide plant growth regulatory substances which induce plant root hair curling and cortical cell division. Although numerous earlier reports had reported the isolation of material from *Rhizobium* cultures that could affect leguminous roots, the first indication of the involvement of the *nod* genes in this process came from the work of van Brussel and colleagues (Zaat et al, 1987; van Brussel et al, 1986; 1990). These authors reported that sterile culture supernatants from *R. leguminosarum* bv. *viciae* cultures would elicit a thick and short root (TSR) phenotype on seedlings of common vetch (*Vicia sativa* subsp. *nigra*). This work also showed that the *nodD* and *nodABC* genes were essential for this effect and that inducers of the *nod* genes must be present. Subsequently, Schmidt et al (1988) reported that culture supernatants from luteolin-induced *R. meliloti* cells would induce cell division in soybean and alfalfa protoplasts. Banfalvi and Kondorosi (1989) reported that expression of the *nodABC* genes of *R. meliloti* in *E. coli* resulted in the release of substances into the culture supernatant which could induce root hair curling on alfalfa and a few other plants. The compounds released by *Rhizobium* appeared to be specific for their host plant. Faucher et al (1988, 1989) reported that, although culture supernatants from *R. meliloti* cells would curl root hairs on a wide variety of plants, the presence of the *nodH* gene rendered the supernatants specific for alfalfa. In addition, mutants in the *nodQ* gene produced compounds which could curl the root hairs of vetch and alfalfa (Banfalvi and Kondorosi, 1989; Cervantes et al, 1989). At least three bioactive compounds were isolated by HPLC from culture supernatants of *R. meliloti* and when added separately could induce root hair curling or when added together could induce significant root cortical cell division to yield an empty, nodule-like structure (Faucher et al, 1989). The formation of infection threads was not detected.

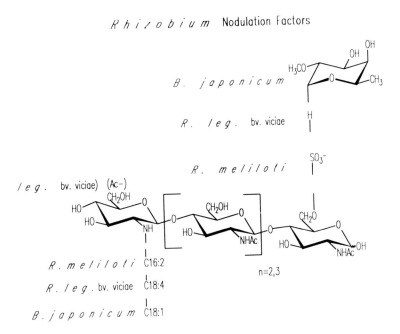

Figure 1. Structures of the nodulation (Nod) factors from B. japonicum, R. meliloti, *and* R. leguminosarum bv. viciae.

The structures of the compounds produced by *R. meliloti* have been determined (Figure 1). The first of these compounds, NodRm-1 is a N-acyl tri N-acetyl ß-1,4-D-glucosamine tetrasaccharide, bearing a sulfate group on C-6 of the reducing sugar (Lerouge et al, 1990a; Truchet et al, 1991). The second compound isolated, NodRm-2, is identical to NodRm-1, but lacks the sulfate substituent. It now seems likely that the NodH, NodP, and NodQ proteins are involved in the conversion of NodRm-2 to NodRm-1 via a sulfation reaction (Lerouge et al, 1990a; Schwedock and Long, 1990; Truchet et al, 1991). This modification is apparently important to the specificity of this compound for *Medicago*. More recently, Roche et al (1991) reported the isolation of additional Nod factors from *R. meliloti*, including a pentasaccharide, and Schultze et al (1992) have reported trisaccharide forms; however, all of these compounds are similar to NodRm-1. A family of Nod factor metabolites has also been isolated from induced culture supernatants of *R. leguminosarum* bv. *viciae* (Figure 1; Spaink et al, 1991a, b). The *R. leguminosarum* bv. *viciae* Nod factor is a pentasaccharide of N-acetylglucosamine modified at the non-reducing end with a polyunsaturated fatty acid and an acetyl group. Analysis of the compounds made by mutants of *R. leguminosarum* suggests strongly that NodFE are involved in production of the fatty acid substituent and NodL is involved in the acetylation reaction. Apparently, these modifications are crucial for host specificity.

The Nod factors either in the crude or purified state have been shown to have a variety of biological activities: 1. They can inhibit root growth (the TSR phenotype) (van Brussel et al, 1986; Zaat et al, 1987). 2. They can induce plant root hair deformation (Had[+] phenotype; Lerouge et al, 1990a, b; Banfalvi and Kondorosi, 1989). 3. They can induce cortical cell division (Coi[+] phenotype; Lerouge et al, 1990a, b; Truchet et al, 1991). 4. They can induce nodule formation (Nod[+] phenotype; Lerouge et al, 1990b; Spaink et al, 1991a; Truchet et al, 1991). 5. They can induce increased exudation of flavonoids, the inducers of *nod* gene expression (the INI, increased nod gene induction, phenotype; van Brussel et al, 1990; Recourt et al, 1991). These biological effects of the isolated Nod factor mirror various aspects of the bacterial infection process summarized above. Therefore, it now seems likely that these plant responses to the bacteria are due to interaction with the Nod factor(s).

The fact that an oligosaccharide is involved in the induction of nodule formation has precedence in the literature. In fact, a class of molecules called oligosaccharins (oligosaccharides with regulatory properties) has been added to the family of plant signal molecules (Albersheim and Darvill, 1985; Albersheim et al, 1983; Hahn and Cheong, 1991). These oligosaccharins have been characterized from fungal, plant and microbial glycoconjugates (Lerouge et al, 1990a; Hahn et al, 1981; York et al, 1984; Nothnagel et al, 1983; McDougall and Fry, 1988; Barber et al, 1989). Various oligosaccharins have been shown to activate the plant defense response during infection (Darvill and Albersheim,

1984; Hahn et al, 1989; Ryan, 1987), to regulate plant hormone responses (York et al, 1984; McDougall and Fry, 1988, 1989a,b), and affect plant development (Eberhard et al, 1989). Therefore, it is perhaps not surprising that a *Rhizobium* oligosaccharin, the Nod factor, is involved in determining symbiosis since this process requires cell-cell recognition, regulation of host defenses and cellular differentiation processes.

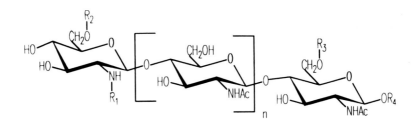

B. japonicum Nodulation Factors

R1	R2	R3	R4	n	M+H+	Strain
C18:1	H	2-O-Me-Fuc	H	3	1416	110 (I)
						135 (I)
						Rx18E (II)
C18:1	Ac	2-O-Me-Fuc	H	3	1458	135
						Rx18E
C16:0	H	2-O-Me-Fuc	H	3	1390	135
C16:0	Ac	2-O-Me-Fuc	H	3	1432	135
C16:1	H	2-O-Me-Fuc	H	3	1388	135
C18:1,Me	Ac	2-O-Me-Fuc	H	3	1472	Rx18E
C18:1	H	Fuc	Gro	2	1273	Rx18E
C18:1	Ac	Fuc	Gro	2	1315	Rx18E
C18:1,Me	Ac	Fuc	Gro	2	1329	Rx18E

Figure 2. The structures of the nodulation factors purified from B. japonicum *strains USDA110 (110), USDA135 (135), and USDA61 (Rx18E).*

Comparison of the Nod factors with known oligosaccharins shows they are most closely related to the biologically active chitin oligomers. In fact, removal of the sulfate and fatty acyl group from the *R. meliloti* NodRm-1 compound would be expected to produce a chitin tetrasaccharide which is known to be very active in eliciting a potent defense response (e.g. lignification, Barber et al, 1989). If one views a *Rhizobium*-legume symbiosis as a modified pathogen-plant interaction, it is tempting to speculate that host-specific *nod* genes, and possibly the common *nod* genes, are actually analogous to bacterial avirulence genes which modify, in a host-specific manner, a chitin-like molecule that would otherwise be recognized by the plant resistance gene product (see Genotype-specific nodulation below).

Lipo-oligosaccharide signals produced by *Bradyrhizobium japonicum.*

Recently, we have been successful in elucidating the structures of the Nod metabolites produced by *B. japonicum* strains USDA135, USDA110 (Sanjuan et al, 1992; Carlson et al, in preparation) and USDA61 (Carlson et al, in preparation). The *B. japonicum* Nod factors are pentasaccharides of N-acetylglucosamine (Figure 2) with a 2-O-methylfucose substituent on the reducing-end sugar.

The non-reducing end is substituted with either an 18:1, 16:0, or 16:1 fatty acid. The non-reducing end may or may not be acetylated. Unlike the Nod factor from *R. meliloti*, none of the compounds isolated from *R. leguminosarum* or *B. japonicum* contain sulfate (Spaink et al, 1992). Comparative analysis of the structure of the Nod factors from the three organisms reveals similarities in that all three are chitin oligomers. Indeed, to our knowledge, this is the only case in prokaryotes where chitin is synthesized. The specificity of these compounds apparently is due to the specific modifications of each structure. These modifications can either be at the reducing end (e.g., *R. meliloti* or *B. japonicum*) or the non-reducing end (e.g., *R. leguminosarum* bv. *viciae*). It may be that these specific modifications tailor the molecule to interact with a specific plant receptor.

B. japonicum strains have been subdivided into two taxonomic subgroups (I and II) based on a number of characteristics (reviewed in Elkan, 1992). In general, group II strains (i.e., cowpea miscellaneous group) have a broader host range than group I strains. Therefore, it was of interest to chemically characterize the Nod factors produced by a group II *B. japonicum* strain. Figure 2 shows the results of our analysis of the Nod metabolites produced by strain Rx18E (a derivative of USDA61, Stokkermans et al, 1992). As in the case for USDA110 and USDA135, we find pentasaccharides of N-acetylglucosamine modified by 2-O-methylfucose. However, the lipo-oligosaccharides signals produced by this strain possess three unique substituents. In two cases, the R1 position (Figure 2) possesses a N-methyl group, in addition to the acyl group. Such methylation has recently been reported in the Nod metabolites produced by *Rhizobium* sp. NGR234 (Price et al, 1992). This strain has the broadest host range of any known *Rhizobium*. Therefore, this substitution may be important for extension of host range. Furthermore, for the first time, tetrasaccharide forms of the nodulation factors were isolated. These tetrasaccharide are unusual in two regards; one, they are substituted with fucose, not 2-O-methylfucose and two, the reducing hydroxyl residue is blocked by a glycerol substituent. We have subsequently found that such glycerol containing Nod metabolites are also produced by strain USDA110 (Luka et al, unpubl.). These unusual tetrasaccharide Nod factors may be end products of metabolism that have specific biological functions. However, it is also possible that they represent intermediates in the biosynthesis of the Nod metabolites. These possibilities are currently under investigation.

Genotype-specific nodulation genes

A number of field studies examining *Rhizobium* ecology have noted that specific soybean cultivars appear to be preferentially nodulated by certain *B. japonicum* strains (reviewed in Stacey and Brill, 1982; Bottomley, 1991). These studies have pointed out the potential importance of this selectivity in explaining competition between *B. japonicum* strains for nodulation. In a few cases, the genetic basis of plant selection of particular strains has been studied. For example, soybeans containing the dominant *Rj4* allele restrict nodulation of *B. japonicum* strain USDA61 (Vest and Caldwell, 1972; Devine et al, 1990) and some serogroup 123 strains (Sadowsky and Cregan, 1992). However, the molecular determinants of the bacteria that control this restriction have not been characterized.

Competition between improved inoculant strains and indigenous soil rhizobia is recognized as a major obstacle to the improvement of soybean productivity by inoculation. For example, in soybean production areas of the midwestern U.S., members of *B. japonicum* serocluster 123 are the dominant indigenous competitors for soybean nodulation (Ellis et al, 1984; Ham et al, 1971; Moawad et al, 1984). These strains are often less efficient at fixing nitrogen than commercially available strains and therefore a method to supplant these indigenous, competitive strains within the nodules could have agronomic benefits. Cregan and colleagues at the USDA Beltsville Laboratory of Nitrogen Fixation initiated a research program to identify soybean genotypes which would exclude nodulation by members of the 123 serocluster and therefore could theoretically favor nodulation by the competing inoculant strain. Cregan and Keyser (1986) identified several soybean genotypes which restricted nodulation and reduced the competitiveness of *B. japonicum* strain USDA123. These genotypes were nodulated normally by other, non-123 serogroup strains (e.g., USDA110). Two of the genotypes, Plant Introduction 377578 and PI 371607, were subsequently shown to differentially restrict nodulation by 20 different serocluster 123 isolates (Keyser and Cregan, 1987). Subsequent work has found other soybean genotypes which will restrict members of serocluster 123 or other strains in serogroups 127 or 129 (Cregan et al, 1989a, b; Cregan and Sadowsky, personal communication). In the case of nodulation restriction of serocluster 123 strains and those of serogroup 127 or 129, a single, dominant plant gene appears to be involved in each case (Cregan, personal communication).

Sadowsky et al (1991) identified a single, dominant bacterial gene which is essential for nodulation of the serocluster 123 restrictive soybean genotypes. This gene was termed *nolA* and is an example of a genotype-specific nodulation gene (GSN) that determines infection of specific plant genotypes within a given legume species. NolA has been proposed to be a DNA binding, transcriptional regulatory protein. The *nolA* gene of *B. japonicum* appears to represent the bacterial counterpart of a gene-for-gene interaction system involved in soybean nodulation. Recent data suggest that NolA acts to inhibit the synthesis of the Nod factors (Sanjuan et al, unpubl. results). This suggests that the levels of nodulation signals and/or production of specific signals may control genotype specific nodulation.

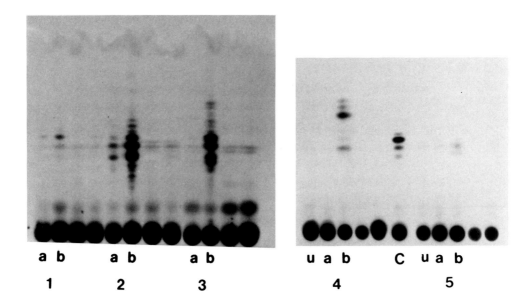

Figure 3. Thin-layer chromatography of nodulation metabolites produced by different wild-type Bradyrhizobium *strains (see Spaink et al, 1992, for methods). a= induced by the addition of 2 µM genistein; b= induced with soybean seed extract; u= no inducer added; C= control lane (i.e., Nod metabolites produced by* B. japonicum *strain USDA110). Strains tested: lane 1=* B. japonicum *strain USDA110; 2= strain USDA135; 3= strain 61A101c; 4=* Bradyrhizobium *sp. (cowpea) strain 8A11A;* Bradyrhizobium *sp. (cowpea) strain 150B1.*

GSN genes have been reported in a few other cases but have not been extensively studied. The first report of such a gene was in *R. leguminosarum* bv. *viciae* strain TOM where a single gene, *nodX*, was identified which allowed this strain to nodulate Afghanistan pea (Lie, 1978; Hombrecher et al, 1984; Gotz et al, 1985; Davis et al, 1988). *nodX* is induced by pea root-exudate or by the flavones eriodictyol and hesperitin. Studies of genetic crosses between European genotypes and Afghanistan pea revealed that nodulation restriction of bv. *viciae* strains was determined by a single, recessive plant gene (Holl, 1975). Therefore, a gene-for-gene interaction appears to exist in which a single, dominant gene in *R. leguminosarum* bv. *viciae* strain TOM determines nodulation of Afghanistan pea as controlled by a single, recessive plant gene (Djordjevic et al, 1987). The biochemical function of NodX is unknown.

In a different study, Pueppke and colleagues found that *R. fredii* strain USDA257 nodulates soybean cultivar Peking normally, but cannot nodulate cultivar McCall (Heron and Pueppke, 1984; 1987). Heron et al (1989) isolated 5 mutants of strain USDA257 in which the host range was extended to include McCall. These mutants are interesting in that they appear to identify bacterial genes that act negatively to restrict nodulation on specific genotypes of soybean.

As mentioned above, soybean lines carrying the Rj4 allele specifically restrict nodulation by *B. japonicum* strain USDA61 (Vest and Caldwell, 1972) and some members of the USDA123 serogroup (Sadowsky and Cregan, 1992). Recently, Stokkermans et al (1992) reported the isolation of mutants of strain USDA61 that could overcome the nodulation restriction of Rj4 soybean. When the profile of Nod metabolites produced by these mutant strains were compared to those of the wild type or other nodulation restricted mutants, clear differences were apparent (Stokkermans et al, 1992). Mutant strains able to nodulate Rj4 soybean appear to produce a profile of Nod metabolites that more closely resembled that of strain USDA110, a strain that is not restricted for nodulation on Rj4 soybean. These data and those obtained by analyzing NolA function (see above) suggest a working hypothesis to explain genotype specific nodulation and, perhaps, the role of Nod metabolites in host specificity. The data suggest that different *Rhizobium* strains produce their own individual profile of Nod metabolites that act on the host plant. Each plant cultivar may react to specific Nod metabolites in a different way. Some Nod metabolites may be inhibitory, preventing nodulation by the specific strain that produced them, resulting in genotype specific nodulation restriction. Therefore, the repertoire of Nod factors produced by a specific strain (as defined by the Nod/Nol functions of that strain) interacts in a complex way with the plant genotype to affect an efficient or inefficient interaction. The important plant functions involved may be specific receptor molecules that interact with the Nod factors and elicit a cellular response leading to nodulation. At this point, plant Nod factor receptors have yet to be identified and many other aspects of this hypothesis remain to be confirmed. However, as shown in Figure 3, when one analyzes the types of Nod factors produced by different *B. japonicum* strains, it is clear that each strain produces a unique profile of metabolites. These results suggest that there is certainly the potential for the plant to recognize different wild-type strains based on the profile of signals produced. This recognition could lead to specific strain selection and be an important determinant in competition between strains for successful nodulation.

Summary

Recent evidence indicates that rhizobia signal to their plant hosts via novel lipo-chitosan molecules. These substances represent a new class of plant growth regulators within the general group of oligosaccharins. These Nod factors appear to be essential for the induction of the early events in nodule formation (e.g., root hair curling and meristem formation) and may be important in other stages. Data suggest that the specific modification of the general chitosan structure of the Nod factor is important for determination of host and genotype specific nodulation. This specificity may play a role in interstrain competition for nodulation.

Several important questions remain to be addressed with regard to Nod factors. One, how are these factors synthesized? For example, are the glycerol-containing metabolites isolated from *B. japonicum* intermediates in biosynthesis? Two, how does the plant recognize these metabolites? Are there plant receptors for these molecules and what is the nature of the cellular signaling pathway? Three, how did this unprecedented (i.e., chitin was not previously known to be produced in prokaryotes) signaling system develop? With regard to this last question, Benhamou and Asselin (1989) presented data that chitin derivatives occur in secondary plant cell walls in a number of plant species. Recently, Spaink et al (1993) presented data suggesting that chitinase-sensitive, lipophilic molecules are present in *Lathyrus* plants. Therefore, one possibility is that the Nod factors resemble previously undiscovered chitin-like regulatory molecules normally present in plants. If this where the case, then the lipo-oligosaccharide nodulation signals would join the list of other plant growth regulators (e.g., cytokinin, auxin, etc.) known to be produced by bacteria that interact intimately with plants.

References

Albersheim, P., Darvill, A.G., McNeil, M., Valent, B.S., Sharp, J.K., Nothnagel, E.A., Davis, K.R., Yamazaki, N., Gollin, D.J., York, W.S., Dudman, W.F., Darvill, J.E. & Dell, A. (1983) In: *Structure and function of plant genomes* (Ciferri, O. & Dure, L. III, eds.) pp 293-312, Plenum Publ. Corp., New York, N.Y.

Albersheim, P. & Darvill, A.G. (1985) *Sci. American* **253**, 58-64.

Banfalvi, Z. & Kondorosi, A. (1989) *Plant Mol. Biol.* **13**, 1-12.

Barber, M.S., Bertram, R.E. & Ride, J.P. (1989) *Physiol. Mol. Plant Pathol.* **34**, 3-12.

Benhamou, N. & Asselin, A. (1989) *Biol. Cell* **67**, 341-350.

Bottomley, P. (1991) In: *Biological Nitrogen Fixation* (Stacey, G., Burris, R.H. & Evans, H.J., eds.) pp 293-348, Chapman and Hall, New York.

Cervantes, E., Sharma, S.B., Maillet, F., Vasse, J., Truchet, G. & Rosenberg, C. (1989) *Mol. Microbiol.* **3**, 745-755.

Cregan, P.B., Keyser, H.H. & Sadowsky, M.J. (1989a) *Appl. Env. Microbiol.* **55**, 2532-2536.

Cregan, P.B., Keyser, H.H. & Sadowsky, M.J. (1989b) *Crop Science* **29**, 307-312.

Cregan, P.B. & Keyser, H.H. (1986) *Crop Science* **26**, 911-916.

Darvill, A.G. & Albersheim, P. (1984) *Ann. Rev. Plant Physiol.* **35**, 243-275.

Davis, E.O., Evans, I.J. & Johnston, A.W.B. (1988) *Mol. Gen. Genet.* **212**, 531-535.

Devine, T.E., Kuykendall, L.D. & O'Neill, J.J. (1990) *Theor. Appl. Genet.* **80**, 33-37.

Djordjevic, M.A., Schofield, P.R. & Rolfe, B.G. (1985) *Mol. Gen. Genet.* **200**, 463-471.

Djordjevic, M.A., Gabriel, D.W. & Rolfe, B.G. (1987) *Ann. Rev. Phytopathol.* **25**, 145-168.

Downie, J.A., Hombrecher, G., Ma, Q.-S., Knight, C.D., Wells, B. & Johnston, A.W.B. (1983) *Mol. Gen. Genet.* **190**, 359-365.

Eberhard, S., Doubrava, N., Marfa, V., Mohnen, D., Southwick, A., Darvill, A. & Albersheim, P. (1989) *Plant Cell* **1**, 747-755.

Elkan, G.H. (1992) *Can. J. Microbiol.* **38**, 446-450.

Ellis, W.R., Ham, G.E. & Schmidt, E.L. (1984) *Agron. Journal* **76**, 573-576.

Faucher, C., Maillet, F., Vasse, J., Rosenberg, C., van Brussel, A.A.N., Truchet, G. & Denarie, J. (1988) *J. Bacteriol.* **170**, 5489-5499.

Faucher, C., Camut, S., Denarie, J. & Truchet, G. (1989) *Mol. Plant-Microbe Int.* **2**, 291-300.

Gloudemans, T. & Bisseling, T. (1989) *Plant Sci.* **65**, 1-14.

Gotz, R., Evans, I.J., Downie, J.A. & Johnston, A.W.B. (1985) *Mol. Gen. Genet.* **201**, 296-300.

Govers, G., Nap, J.-P., Van Kammen, A. & Bisseling, T. (1987) *Plant Physiol. Biochem.* **25**, 309-322.

Hahn, M.G., Darvill, A.G. & Albersheim, P. (1981) *Plant Physiol.* **68**, 1161-1169.

Hahn, M.G., Bucheli, P., Cervone, F., Doares, S.H., O'Neill, R.A., Darvill, A. & Albersheim, P. (1989) In: *Plant-Microbe Interactions. Molecular and genetic perspectives.* Vol. 3 (Kosuge, T. & Nester, E.W., eds.) pp 131-181, McGraw Hill Publ. Ço., New York, N.Y..

Hahn, M.G. & Cheong, J.-J. (1991) in: *Advances in Molecular Genetics of Plant-Microbe Interactions* (Hennecke, H. & Verma, D.P.S., eds.) pp 403-420, Kluwer Acad. Publ., Dordrecht, The Netherlands.

Ham, G.E., Caldwell, V.B. & Johnson, H.W. (1971) *Agron. Journal* **63**, 301-303.

Heron, D.S., Ersek, T., Krishnan, H.B. & Pueppke, S.G. (1989) *Mol. Plant-Microbe Int.* **2**, 4-10.

Heron, D.S. & Pueppke, S.G. (1984) *J. Bacteriol.* **160**, 1061-1066.

Heron, D.S. & Pueppke, S.G. (1987) *Plant Physiol.* **84**, 1391-1396.

Holl, F.B. (1975) *Euphytica* **24**, 767-770.

Hombrecher, G., Gotz, R., Dibb, N.J., Downie, J.A., Johnston, A.W.B. & Brewin, N.J. (1984) *Mol. Gen. Genet.* **194**, 293-2980.

Horvath, B., Bachem, C.W.B., Schell, J. & Kondorosi, A. (1987) *EMBO J.* **6**, 841-848.

Keyser, H.H. & Cregan, P.B. (1987) *Appl. Env. Microbiol.* **53**, 2631-2635.

Kondorosi, A., Kondorosi, E., Pankhurst, C.E., Broughton, W.J. & Banfalvi, Z. (1984) *Mol. Gen. Genet.* **193**, 445-452.

Kondorosi, A., Horvath, B., Göttfert, M., Putnoky, P., Rostas, K., Gyorgypal, Z., Kondorosi, E., Torok, I., Bachem, C.W.B., John, M., Schmidt, J. & Schell, J. (1985) In: *Nitrogen Fixation Research Progress* (Evans, H.J., Bottomley, P.J. & Newton, W.E., eds.) pp 73, Marinus Nijhoff, Dordrecht, The Netherlands.

Lerouge, P., Roche, P., Faucher, C., Maillet, F., Truchet, G., Prome, J.C. & Dénarié, J. (1990a) *Nature* **344**, 781-784.

Lerouge, P., Roche, P., Prome, J.-C., Faucher, C., Vasse, J., Maillet, F., Camut, S., de Billy, F., Barker, D.G., Denarie, J. & Truchet, G. (1990b) In: *Nitrogen Fixation: Achievements and Objectives* (Gresshoff, P.M., Roth, L.E., Stacey, G. & Newton, W.E., eds.) pp 177-186, Chapman and Hall, New York.

Lie, T.A. (1978) *Ann. Appl. Biol.* **88**, 462-465.

Long, S.R., Buikema, W. & Ausubel, F.M. (1982) *Nature* **298**, 485-488.

McDougall, G.J. & Fry, S.C. (1988) *Planta* **175**, 412-416.

McDougall, G.J. & Fry, S.C. (1989a) *J. Exp. Bot.* **40**, 233-238.

McDougall, G.J. & Fry, S.C. (1989b) *Plant Physiol.* **89**, 883-887.

Moawad, H.A., Ellis, W.R. & Schmidt, E.L. (1984) *Appl. Env. Microbiol.* **47**, 607-612.

Mulligan, J.T. & Long, S.R. (1985) *Proc. Natl. Acad. Sci. (USA)* **82**, 6609-6613.

Nothnagel, E.A., McNeil, M., Albersheim, P. & Dell, A. (1983) *Plant Physiol.* **71**, 916-926.

Price, N.P.J., Relic, B., Talmont, F., Lewin, A., Prome, d., Pueppke, S.G., Maillet, F., Denarie, J., Prome, J.-C. & Broughton, W.J. (1992) *Mol. Microbiol.* **6**, in press.

Recourt, K., Schripsema, J., Kijne, J.W., van Brussel, A.A.N. & Lugtenberg, B.J.J. (1991) *Plant*

Mol. Biol. **16**, 841-852.

Roche, P., Lerouge, P., Ponthus, C. & Prome, J.C. (1991) *J. Biol. Chem.* **266**, 10933-10940.

Rossen, L., Shearman, C.A., Johnston, A.W.B. & Downie, J.A. (1985) *EMBO J.* **4**, 3369-3373.

Roth, L.E., Jeon, K. & Stacey, G. (1988) In: *Molecular Genetics of Plant-Microbe Interactions* (Palacios, R. & Verma, D.P.S., eds.) pp 220-225, The American Phytopathological Society, St. Paul, MN.

Ryan, C.A. (1987) *Ann. Rev. Cell* **3**, 295-317.

Sadowsky, M.J., Cregan, P.B., Göttfert, M., Sharma, A., Gerhold, D., Rodriguez-Quinones, F., Keyser, H.H., Hennecke, H. & Stacey, G. (1991) *Proc. Natl. Acad. Sci. (USA)* **88**, 637-641.

Sadowsky, M.J. & Cregan, P.B. (1992) *Appl. Env. Microbiol.* **58**, 720-723.

Sanjuan, J., Carlson, R.W., Spaink, H.P., Bhat, U.R., Barbour, W.M., Glushka, J. & Stacey, G. (1992) *Proc. Natl. Acad. Sci. (USA)* **89**, 8789-8793.

Schmidt, J., Wingender, R., John, M., Wieneke, U. & Schell, J. (1988) *Proc. Natl. Acad. Sci. (USA)* **85**, 8578-8582.

Schofield, P.R., Ridge, R.W., Rolfe, B.G., Shine, J. & Watson, J.M. (1984) *Plant Mol. Biol.* **3**, 3-11.

Schultze, M., Quicletsire, B., Kondorosi, E., Virelizier, H., Glushka, J.N., Endre, G., Gero, S.D. & Kondorosi, A. (1992) *Proc. Natl. Acad. Sci. (USA)* **89**, 192-196.

Schwedock, J. & Long, S.R. (1990) *Nature* **348**, 644-646.

Scott, K.F., Hughes, J., Gresshoff, P.M., Beringer, J., Rolfe, B.G. and Shine, J. (1982). *J. Molec. Appl. Genet.*, **1**: 315-326.

Shearman, C.A., Rossen, L., Johnston, A.W.B. & Downie, J.A. (1986) *EMBO J.* **5**, 647-652.

Spaink, H.P., Geiger, O., Sheeley, D.M., van Brussel, A.A.N., York, W.S., Reinhold, V.N., Lugtenberg, B.J.J. & Kennedy, E.P. (1991a) In: *Advances in Molecular Genetics of Plant-Microbe Interactions.* Vol. 1 (Hennecke, H. & Verma, D.P.S., Eds.) pp 142-149, Kluwer Academic Publ., Dordrecht, The Netherlands.

Spaink, H.P., Sheeley, D.M., van Brussel, A.A.N., Glushka, J., York, W.S., Tak, T., Geiger, O., Kennedy, E.P., Reinhold, V.N. & Lugtenberg, B.J.J. (1991b) *Nature* **354**, 125-130.

Spaink, H.P., Aarts, A., Stacey, G., Bloemberg, G.V., Lugtenberg, B.J.J. & Kennedy, E.P. (1992) *Mol. Plant-Microbe Int.* **5**, 72-80.

Spaink, H.P., Aarts, A., Bloemberg, G.V., Folch, J., Geiger, O., Schlaman, H.R.M., Thomas-Oates, J.E., Van De Sande, K., Van Spronsen, P., van Brussel, A.A.N., Wijfjes, A.H.M. & Lugtenberg, B.J.J. (1993) In: *Advances in Molecular Genetics of Plant-Microbe Interactions.* Volume 2 (Nester, E. & Verma, D.P.S., Eds.) pp 151-162, Kluwer Academic Publishers, Dordrecht, The Netherlands.

Stacey, G. & Brill, W.J. (1982) in: *Phytopathogenic Prokaryotes*. Vol. I (Mount, M.S. & Lacy, G.H., Eds.) pp 225-247, Academic Press, New York.

Stokkermans, T.J.W., Sanjuan, J., Ruan, X., Stacey, G. & Peters, N.K. (1992) *Mol. Plant-Microbe Int*. **5**, 504-512.

Truchet, G., Roche, P., Lerouge, P., Vasse, J., Camut, S., de Billy, F., Prome, J.-C. & Denarie, J. (1991) *Nature* **351**, 670-673.

van Brussel, A.A.N., Zaat, S.A.J., Canter-Cremers, H.C.J., Wijffelman, C.A., Pees, E., Tak, T. & Lugtenberg, B.J.J. (1986) *J. Bacteriol*. **165**, 517-522.

van Brussel, A.A.N., Recourt, K., Pees, E., Spaink, H.P., Tak, T., Wijffelman, C.A., Kijne, J.W. & Lugtenberg, B.J.J. (1990) *J. Bacteriol*. **172**, 5394-5401.

Verma, D.P.S., Fortin, M.G., Stanley, J., Mauro, V.P., Purohit, S. & Morris, N. (1986) *Plant Mol. Biol*. **7**, 51-61.

Verma, D.P.S. & Long, S.R. (1983) *Int. Rev. Cytol. Suppl*. **14**, 211-245.

Vest, G. & Caldwell, B.E. (1972) *Crop Science* **12**, 692-693.

York, W.S., Darvill, A.G. & Albersheim, P. (1984) *Plant Physiol*. **75**, 295-297.

Zaat, S.A.J., van Brussel, A.A.N., Tak, T., Pees, E. & Lugtenberg, B.J.J. (1987) *J. Bacteriol*. **169**, 3388-3391.

Plant Response to the Microgravity Environment of Space

Ronald L. Schaefer[1], Gary C. Jahns[2], and Debra Reiss-Bubenheir

[1]Lockheed Engineering & Sciences Co. and [2]Space Life Sciences Payloads Office, NASA Ames Research Center, Moffett Field, CA 94035-1000 USA

Introduction

This chapter highlights recent research results on plant responses to the microgravity environment of space. Several detailed reviews that include early plant space research can be found in the literature (Halstead and Dutcher, 1984, 1987; Halstead and Scott, 1984, 1990; Krikorian, 1991; Krikorian and Levine, 1991). Brown and Chapman (1984a) describe "bioengineering" tests with hardware designed to house plant seedlings and some of the unique aspects of conducting research in space. Plant research in space i focused primarily on basic biology: gravity perception and plan movement/responses to various stimuli when the influence of the earth's 1; field is removed. The results from basic biology studies will lead to th eventual production of plants in space to provide food and re-generat oxygen, basic goals of programs such as the Controlled Ecological Life Suppor System (CELSS) of the National Aeronautics and Space Administratio: (NASA), and critical to man's long-term presence in space or in reduce gravity (moon, Mars) environments.

0-8493-8263-7/93/$0.00 + $.50

© 1993 by CRC Press, Inc.

Most of the work reported will be from experiments funded by NASA and carried in various Biosatellites and more recently several Space Transportation System (STS or Space Shuttle) orbiters. Experiments carried in Russian spacecraft such as Soyuz, Cosmos (unmanned biological satellites; described by Souza, 1979), and the space stations Salyut and Mir will also be covered.

Numerous terms are used to describe the microgravity environment of space; however, the generally accepted definitions may be found in Table 1. Most plant studies conducted in space have been performed in microgravity, i.e., at 10^{-6} to $10^{-3}g$, although reference to "zero gravity" and weightlessness will be found in the literature.

Table 1. Gravity definitions

Hypergravity	*Greater than 1g*
Hypogravity	*Less than 1g*
Microgravity	*10^{-6} to $10^{-3}g$*
Zero gravity	*0g, theoretical, cannot be achieved*
Weightlessness	*Implies 0g, but actually is a condition with some finite g force. Found briefly in free fall (Pollard, 1965).*

Gravity is a constant environmental factor with which all living things on earth have had to contend with for more than 3.5 billion years. Biologists have been studying plant response to altered gravity for years. Halstead and Scott (1990) claim "plants unquestionably were the first organisms to indicate a biological role for gravity". The response of plants to 1g is clear: the roots elongate downward in a positive geotropic response, while the shoot grows upward in a negatively geotropic manner. It is difficult to measure the response of plants to microgravity in a 1g environment. Devices, such as a clinostat that provides gravity compensation by rotation around a horizontal axis, have been used to simulate microgravity conditions (Brown et al, 1974 and 1976; Eidesmo et al, 1991; Chapman and Brown, 1979; Chapman et al, 1980; Heathcote and Bircher, 1987; Rasmussen et al, 1989; Shen-Miller et al, 1968), but the microgravity environment of space must be utilized to study the actual responses.

Gravity perception

The general nature of the gravity sensing mechanism in plants has been extensively described (Iverson, 1969; Pickard, 1985), but the precise transduction of sensory inputs into metabolic changes are unknown. Roux (1990) reviewed papers that put forth the hypothesis that calcium ions, together with hormones and other factors, play an important role in plant gravity perception.

Figure 1. Representative mung bean seedlings grown in a Plant Growth Unit for 8 days aboard STS-51F. Note unusual orientation of root growth. Photograph taken about 3 hours after landing with the lid of the Plant Growth Chamber removed (Cowles et al, 1988). Photograph courtesy of NASA, printed with permission.

It is well known, dating back to classic studies (Darwin, 1880; Went, 1932), that the root cap is solely responsible for gravity perception in roots. Iverson (1969) found that when statolith starch was removed from garden cress (*Lepidium sativum*) roots gravity perception was also removed. Iverson and others have shown that the gravity sensing cells, called statocytes, are present in the center of the root cap.

Perbal et al (1986, 1987) studied lentil roots (*Lens culinaris*) grown in the dark for 25 or 35 hours on moistened cellulose sponges during the D-1 Spacelab mission. They found that the statocytes had normal cell polarity (endoplasmic reticulum near the distal wall, nucleus near the proximal wall), but the amyloplasts of the space-exposed roots were located in the center of the cell and had almost no contact with the endoplasmic reticulum. Volkmann et al (1986) similarly found that amyloplasts in garden cress roots flown on the same mission did not sediment on the endoplasmic reticulum as occurs on earth, but instead were randomly distributed.

Early events in geotropism in plant shoots are reviewed by Pickard (1985) who states that the first step in gravity perception results from the movement of amyloplasts through the cytoplasm or from the settling of the amyloplasts to the bottom portion of the cytoplasm. Heathcote (1981) found that gravity perception in mung bean (*Vigna radiata*) hypocotyls occurred by settling of the amyloplasts in the starch sheath (starch sheath is identical to the endodermal ring in mung bean).

Cellular activity

The use of plant cell cultures in space began with the joint US - Russian biological satellite program. Krikorian and Steward (1978) tested carrot (*Daucus carota*) cells under the microgravity conditions experienced aboard Cosmos 782. They found that totipotent somatic cells proceeded through morphogenesis in the dark for 20 days in space. The cells produced embryos comparable to the 1*g* controls aboard the satellite.

Krikorian and O'Connor (1984) conducted karyological observations on roots of three plant species grown in space. Root samples were from Brown and Chapman (1984b) who flew experiments on Shuttle flights STS-2 (4 days) and STS-3 (8 days) and from the STS-3 experiment of Cowles et al (1984, 1989). Krikorian and O'Connor found evidence that cells in sunflower (*Helianthus annuus*) roots subjected to space flight exhibited major changes in their cell division profile, even after as few as 4 days in space. Analyses of several root tips revealed a failure of all the dividing nuclei to arrest in the metaphase stage. Bridge formation was also apparent. Paucity of cell division, and considerable chromosome fragmentation and breakage in space-grown oat (*Avena sativa)* roots were noted by the authors. No aberrations have been found in mung bean roots (*Vigna radiata)*; chromosomes in flight samples were smaller than normally seen on earth, but metaphase stages were present in large enough numbers for comparisons to be made. No major differences were noted except a reduction in total number of dividing cells in microgravity-grown roots. The authors conclude that the roots of each plant species exhibited considerable reduction in initial cell division. This could be due either to the microgravity environment of space and/or space flight conditions (vibration effects from launch or recovery and radiation effects).

The anatomy and fine structure of mung bean and oat roots, also from STS-3, were studied by Slocum et al (1984). Space-grown mungbean roots had normal tissue organization, but their root cap cells were collapsed and degraded in appearance. Roots of microgravity-grown oat seedlings were normal except for cortex cell mitochondria which showed a "swollen" morphology. Lewis and Moore (1990) also compared space-grown roots to those grown on earth and found lower amounts of starch and increased vacuoles in amyloplasts of corn (*Zea mays*) roots germinated aboard STS 61-C.

Circadian rhythm

The concept of the endogenously driven circadian oscillator was investigated on two shuttle flights by growing the filamentous fungus, *Neurospora crassa* (pink bread mold) (Ferraro et al, 1989). *Neurospora* expresses a diurnal rhythm in asexual spore formation (conidiation). Findings from the first flight of this experiment (STS-9) and from ground-based centrifugation studies suggested gravity sensing input to the pacemaker; however, the results were not conclusively supportive of endogenous generation of biological rhythms (some rhythm damping occurred) primarily due to the hypergravity experienced during launch. This hypothesis was confirmed by ground-based studies. A second experiment was then conducted in January 1990 as part of the STS-32 mission, the first longer duration flight (11 days). The middeck experimental package was redesigned to counteract the Zeitgeber (circadian rhythm or phase cue) of launch by exposing some of the cultures to a light pulse (successful in ground-based studies) early in flight while others were light-filtered. The astronauts exposed the cultures to the light and marked mycelia growth fronts. Results from STS-32 demonstrated that the conidiation rhythm persisted in space with minor fluctuations in amplitude and period, suggesting the presence of an endogenously driven circadian oscillator that is modified by environmental inputs. The growth rate of the fungus was also increased, suggesting an increased metabolic rate during space flight (Ferraro et al, 1990).

Plant orientation

Merkys and Laurinavicius (1990) found that initial germination of peas (*Pisum sativum*) in microgravity was normal -- the axial organs of the seedlings grew in the direction determined by their location in the embryo -- in 2-day experiments aboard Soyuz-12 and Soyuz-13 in 1973. Normal germination was reported in earlier experiments performed with pepper (*Capsicum annuum*) and wheat aboard Biosatellite II in 1966 (Saunders, 1971), with mungbean and oat (*Avena sativa*) on STS-51F (Cowles et al, 1988) and more recently with garden cress aboard D-1 in 1985 (Volkmann et al, 1986). However, Krikorian and O'Connor (1984) reported that several sunflower

(*Helianthus annuus*) seedlings flown as part of Brown and Chapman's (1984a) STS-2 Shuttle flight had roots growing out of the soil or growing at the soil level and barely penetrating the growing media. This condition was also noted in mungbean and oat by Cowles et al (1988) (Figure 1).

Plant growth

Endogenous plant hormones regulate growth: to date, only short-term studies have been conducted in space aboard the Shuttle. Schulze et al (1992) found mass and hormonal content of corn (*Zea mays*) seedlings grown in the dark on moistened filter paper for 5 days aboard STS-34 were not significantly different from seedlings grown on earth. The tissues of the space-grown plants appeared normal -- the seedlings differed only in the lack of orientation of the shoots and roots (Figure 2).

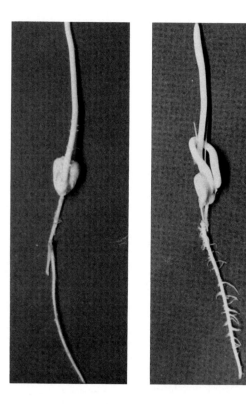

Figure 2. (Left) A representative corn seedling grown on the ground under environmental parameters as close as possible to those of space grown seedlings. (Right) A representative seedling grown at 3 x 10^{-3} g in space. About 50% of the seedlings showed a loop such as shown in this figure or other abnormalities that included the root and shoot growing parallel to each other. Photographs courtesy Dr. Robert S. Bandurski, Michigan State University. Reproduced with permission.

There appeared to be no difference in overall metabolic activity or cell wall thickness. In addition, the content of free and total indole-3-acetic acid were statistically similar in both earth and space grown seedlings. These results imply that plants can be grown successfully for CELSS provided an orienting stimulus (light or electrical) is applied.

Levine and Krikorian (1991) found no difference between shoot growth of ground control daylily (*Hemerocallis*) or the dicot *Haplopappus gracilis* and plants grown for 5 days in a Plant Growth Unit (PGU) aboard STS-29. Enhanced root growth was observed for the space-exposed *Haplopappus* tissue culture plantlets. Pre-existing root primordia emerged more quickly under the microgravity conditions of space.

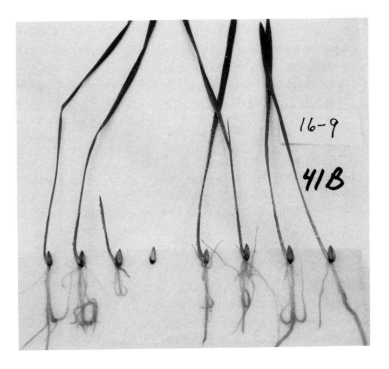

Figure 3. Representative oat seedlings grown in a Plant Growth Unit for 8 days aboard STS-51F. Note unusual circular pattern of root growth in 1 plant. Photograph taken about 3 hours after landing (Cowles et al, 1988). Photograph courtesy of NASA, printed with permission.

Lignin is one of the major structural components found in higher plants and is critical in the support of the vertical growth habit of plants in a gravity environment. Since the appearance of lignin coincides with the evolutionary development of terrestrial plants it has been postulated that lignin deposition might be influenced by gravity (Siegel and Siegel, 1981). Cowles et al (1989) investigated this hypothesis on two 8-day Shuttle missions, STS-3 and STS-51 F. Lignin content and key enzymes in the lignin pathway were measured in three plant species, mungbean (*Vigna radiata*), oat (*Avena sativa*, see Figure 3) and pine (*Pinus elliotti*). Significant reductions in lignin content were found in all three species and ranged from -6 to -24% of the control groups. Phenylalanine ammonia lyase (PAL) and peroxidase enzyme activities were reduced by approximately 20% in the flight seedling relative to 1g controls. The authors suggest that while these reductions in lignin content were significant, longer duration experiments examining lignin content in mature plants grown under higher light levels and at various g levels will be needed to clearly understand the role of gravity in the lignification process.

Circumnutation

Charles Darwin (1875, 1880) reported that plant organs revolve in an elliptical pattern around an axis. This pattern was referred to as circumnutation. Brown and Chapman (1984b) found evidence of circumnutation in 4-day old sunflower (*Helianthus annuus*) seedlings grown in the microgravity environment of Spacelab 1. They concluded from this experiment that circumnutation cannot simply be a response to the earth's 1g field. As of this writing, results have not been fully analyzed from experiments conducted by the same group with wheat (*Triticum aestivum*) on the First International Microgravity Laboratory (IML-1) mission that flew in 1992. Early indications are that circumnutation was again seen in the microgravity environment of space (Heathcote, 1992).

Reproduction and development

It is important to determine the effects of microgravity or fractional gravity on the reproduction and development of plants over multiple generations. Because these experiments require long periods of time in space, only the Russians have performed seed-to-seed experiments aboard their space stations Salyut and Mir.

Merkys and Laurinavicius (1990) reported on the results from several experiments. In one experiment, *Arabidopsis* was grown for 69 days aboard Salyut-7. Five of the seven space-grown plants had 22 normal pods containing 200 ripe seeds. Two plants produced 11 seedless pods. In comparison, eight plants grown in the same apparatus on the ground produced 34 normal pods.

Lack of pod maturity was due, at least in part, to delayed flowering of the space-grown plants. At 44 days after germination plants grown in microgravity were at the same reproductive stage of 36-day old earth-grown plants. The number of seeds per pod was similar for both sets of plants, but viability of those seeds was reduced in the space-grown plants -- although some seeds did germinate and become normal plants.

The effects of microgravity and radiation on *Tradescantia* budded cuttings irradiated aboard Biosatellite II for 2 days were studied by Sparrow et al (1968). They found higher rates of pollen abortion and stamen-hair stunting, suggesting a synergistic effect between microgravity and radiation.

Striking results were obtained when *Arabidopsis* was grown aboard Mir: development of reproductive organs took 20 days longer for plants grown in microgravity when compared to their ground controls (Merkys and Laurinavicius, 1990). However, environmental conditions (temperature, light, humidity) for both space-grown and earth-grown may not have been equivalent.

Conclusions

The field of space biology is still in its infancy and this is especially true for plant biology research. Most of the research is not repeated on follow-up flights, making rigorous statistical analysis difficult. In many cases, data analysis is hampered by inability to duplicate environmental conditions experienced by specimens in space-flown hardware. The area of adequate controls is an important consideration, and as Sharp and Vernikos (1992) point out, factors such as delays between loading and lift-off, lift-off forces, and re-entry forces need to be adequately addressed in controls.

The reader will notice that the duration of the research reported, with the exception of the Russian studies on Salyut and Mir, has been at most a few days. Opportunities to fly are infrequent and Shuttle missions are just now beginning to exceed 10 days. The U.S./International Space Station Freedom is one bright prospect that promises to be an important long-duration platform that launches plant biology, indeed, all space life sciences research of tomorrow. Long-term experiments will be accommodated on Freedom and on the Russian Mir to study plant metabolism, photosynthetic efficiency, and nutrient uptake. In short, overall plant growth and developmental studies must be performed since plants are the key to providing food and replenishing oxygen in space -- the primary tenets of a controlled ecological life support system. With regard to the upcoming development of Freedom, Halstead and Krauss (1992) state that "biologists are poised to take advantage of the greater space, the increased power, and especially the long duration of the station for a cascade of innovative experiments in fundamental science which are long overdue. The unique space environment will provide new dimensions for approaching some of the most challenging problems still facing modern biology."

References

Brown, A.H., Chapman, D.K. & Liu, S.W.W. (1974) *BioScience* **24**, 18-520.

Brown, A.H. & Chapman, D.K. (1984a) *Annals of Bot.* **54** (Suppl. 3), 19-31.

Brown, A.H. & Chapman, D.K. (1984b) *Science* **225**, 230-232.

Brown, A.H., Dahl, A.O. & Chapman, D.K. (1976) *Plant Physiol.* **58**, 127-130.

Chapman, D.K., Venditti, A.L. & Brown, A.H. (1980) *Plant Physiol.* **65**, 533-536.

Chapman, D.K. & Brown, A.H. (1979) *Plant & Cell Physiol.* **20**, 473-478.

Cowles, J.R., LeMay, R., Jahns, G., Scheld, W.H. & Peterson, (1989) In: *Plant and Cell Wall Polymers,* Lewis, N.G., and Paice, G.M. (eds.). pp 203-213. Amer. Chem. Soc., Washington, D.C.

Cowles, J.R., LeMay, R. & Jahns, G. (1988) *Astro. Lett. & Communications* **27**, 223-228.

Cowles, J.R., Scheld, H.W., LeMay, R. & Peterson, C. (1984) *Annals of Bot.* **54** (Suppl. 3), 33-48.

Darwin, C.R. (1875) *The Movements and Habits of Climbing Plants.* John Murray, London.

Darwin, C.R. (1880) *The Power of Movements in Plants.* D. Appelton & Co., New York.

Eidesmo, T., Brown, A.H., Chapman, D.K. & Johnsson, A. (1991) *Micrograv. Sci. Technol.* 4:199-206.

Ferraro, J.S., Fuller, C.A. & Sulzman, F.M. (1989) *Adv. Space Res.* **9**, 251-260.

Ferraro, J. S., Sulzman, F.M., Happe-Shelton, D.J., Pence, M.L., Golay, S.R., Ekborg, K.H. & Dorsett, J.A. (1990) *Soc. Neurosci. Abstr.* **16**, 1333.

Halstead, T.W. & Dutcher, F.R. (1984) *Annals of Bot.* **54** (Suppl. 3), 3-18.

Halstead, T.W. & Dutcher, F.R. (1987) *Ann. Rev. Plant Physiol.* **38**, 317-345.

Halstead, T.W. & Krauss, R.W. (1992) In: *AIAA Space Programs and Technologies Conference,* Paper AIAA-92-1342, Amer. Institute of Aeronautics and Astronautics, Washington, D.C.

Halstead, T. & Scott, T.K. (1984) *Plant gravitational and space research. Workshop Summaries III,* Amer. Soc. Plant Physiol. Waverly Press, Baltimore, Maryland.

Halstead, T. & Scott, T.K. (1990) In: *Fundamentals of Space Biology,* Asashima, M. & Malacinski, G.M. (eds.). pp 9-19. Springer-Verlag, New York.

Heathcote, D.G. (1992) Personal communication.

Heathcote, D.G. (1981) *Plant Cell Environ.* **4**, 131-140.

Heathcote, D.G. & Bircher, B.W. (1987) *Planta* **170**, 249-256.

Iverson, T.H. (1969) *Physiol. Plant.* **22**, 1251-1262.

Krikorian, A.D. (1991) *ASGSB Bull.* **4**, 65-72.

Krikorian, A.D. & Levine, H.G. (1991) In: *Plant Physiology: A Treatise, Growth and Development*, Vol. X. pp 491-555. Academic Press, New York.

Krikorian, A.D. & O'Connor, S.A. (1984) *Annals of Bot.* **54** (Suppl. 3), 49-63.

Krikorian, A.D. & Steward, F.C. (1978) *Science* **200**, 67-68.

Levine, H.G. & Krikorian, A.D. (1991) *ASGSB Bull.* **5**:28.

Lewis, M.L. & Moore, R. (1990) *ASGSB Bull.* **4**, 77.

Malacinski, G.M. & Neff, A.W. (1990) In: *Fundamentals of Space Biology*. Asashima, M. & Malacinski, G.M. (eds.). pp 1-7. Springer-Verlag, New York .

Merkys, A.J. & Laurinavicius, R.S. (1990) In: *Fundamentals of Space Biology*, Asashima, M. & Malacinski, G.M. (eds.). pp 69-83. Springer-Verlag, New York.

Perbal, G., Driss-Ecole, D., Salle, G. & Raffin, J. (1986) *Naturwissenschaften* **73**, 444-446.

Perbal, G. & Driss-Ecole, D. (1987) In: *Third European Symposium on Life Sciences Research in Space* - ESA SP-271. pp 91-96. European Space Agency, Paris, France.

Phillips, R.W. & Haddy, F.J. (1992) In: *AIAA Space Programs and Technologies Conference,* Paper AIAA-92-1344, Amer. Institute of Aeronautics and Astronautics, Washington.

Pollard, E.C. (1965) *J. Theoret. Biol.* **8**, 113-123.

Pickard, B.G. (1985) *Ann. Rev. Plant Physiol.* **36**, 55-75.

Rasmussen, O., Baggerud, C. & Iverson, T.-H. (1989) *Physiol. Plant.* **76**, 431-437.

Roux, S.J. (1990) In: *Fundamentals of Space Biology*. Asashima, M. & Malacinski, G.M. (eds.). pp 57-67. Springer-Verlag, New York.

Saunders, J.F., ed. (1971) *The experiments of Biosatellite II.* NASA SP-204, NASA, Scientific and Technical Information Office, Washington, D.C.

Schulze, A., Jensen, P.J., Desrosiers, M., Buta, J.G. & Bandurski, R.S. (1992) *Plant Physiol.* **100**, 692-698.

Sharp, J.C. & Vernikos, J. (1992) In: *AIAA Space Programs and Technologies Conference,* Paper AIAA-92-1343, Amer. Institute of Aeronautics and Astronautics, Washington, D.C.

Shen-Miller, J., Hinchman, R. & Gordon, S.A. (1968) *Plant Physiol.* **43**, 338-344.

Siegel, S.M. & Siegel, S. (1981) In: *Life in the Universe.* J. Billingham (ed.). p 307. MIT Press, Boston, USA.

Slocum, R.D., Gaynor, J.J. & Galston, A.W. (1984) *Annals of Bot.* **54** (Suppl. 3), 65-76.

Souza, K.A. (1979) *BioScience* **29**, 160-167.

Sparrow, A.H., Shairer, L.A. & Marimuthu, K.M. (1968) *BioScience* **18**, 582-590.

Volkmann, D., Behrens, H.M. & Sievers, A. (1986) *Naturwissenschaften* **73**, 438-441.

Went, F.W. (1932) *J. Wiss. Bot.* **39**, 528-557.

Air Pollution
and Plant Gene Expression

Andreas Bahl, Stefan M. Loitsch and Günter Kahl

Plant Molecular Biology Group, Department of Biology, Johann Wolfgang Goethe-Universität, Siesmayerstr. 70, D-W-6000 Frankfurt/M, Germany

Introduction

The last two decades have seen a steady increase in the area of declined forests both in Europe and the United States as well. This alarming phenomenon has been attributed to the effect of air pollutants of anthropogenic origin, that are mostly set free by combustion processes. One of the potentially harmful compounds, sulphur dioxide, has been known as causative agent of classical smoke damage since medieval ages, and its effects on vegetation been explored and described in detail (for a review see: Ziegler, 1975; Koziol, 1984; Darrall, 1989; Saxe, 1991). However, the emission of sulphur dioxide has been successfully reduced during the past decade through the introduction of retention filters in industrial countries. On the contrary, the ever-increasing automobile density has lead to a continuous accumulation of nitrogen oxide (NO_x) and other, secondary air pollutants, collectively coined photooxidants. These compounds arise by complex light-triggered chain reactions, involving radicals, in which primary emission products (e.g. nitric oxide, NO, and hydrocarbons) are converted to secondary pollutants (e.g. nitrogen dioxide, NO_2, ozone, O_3, and peroxyalkyl nitrates, PAN; Rowland et al, 1985). Though the effect of ozone on vegetation has been explored in some detail, and a wealth of data on the action of NO_x accumulated, very little is known about how automobile exhaust affects plants. The gaseous components of this complex mixture are nitrogen oxides, carbon monoxide (CO), carbon dioxide (CO_2), and hydrocarbons (HC), with the main gases being CO_2, CO, and NO_x, in decreasing order. Keeping this complexity in mind, research on the mechanisms of forest decline should focus on both single components of automobile exhaust and the complete complex mixture. The present review portrays our present, rather incomplete knowledge of both aspects and capitalizes on our own experiments designed to unravel the impact of single components and complex mixtures of exhaust gas on plant gene structure and expression.

Mechanisms of Plant Response to Air Pollutants

Overall effects

The impact of air pollutants on vegetation may long go unrecognized, unless the damage becomes visible. For example, ozone usually causes chlorotic and necrotic symptoms (Smith, 1990), and markedly reduces plant growth and crop yield with drawbacks in agriculture and forestry (Reich, 1987). In contrast, the effects of NO_x and automobile exhaust upon vegetation are less clearcut. Nitrogen oxides even seem to have a beneficial effect on growth, though high concentrations of NO_x lead to visible damage and reduction of growth as does an exposure to low NO_x doses over a long period of time (Roberts et al, 1983; Darrall, 1989). High concentrations of automobile exhaust of up to 1000 µl/l induces yellowing and abscission of needles within a short time (Kammerbauer et al, 1986, 1987) Moreover, potted birch trees, placed by the roadside of a busy highway for three to five months, react with reduced leaf and shoot growth and increased leaf abscission (Flueckinger et al, 1988). Lower concentrations of exhaust gas, however, augment the biomass of young apical leaves of *Helianthus annuus*, but tend to reduce it in the older basal leaves (Schaub et al, 1991). The overall effects of air pollutants on plants then are multifarious and depend on both the relative concentration of individual exhaust compounds, the time of exposure and the age and physiological condition of the plant.

Pollutant uptake

Air pollutants are usually taken up through the stomata. However, NO_2 and NO may also be absorbed through the cuticle, especially at night (Saxe and Murali, 1989). NO_x obviously do not influence stomatal conductance (for a review see Darrall, 1989; Saxe, 1991), whereas the effects of ozone on this process are yet contradictory. Thus, long-term fumigations with ozone generally lead to a closure of stomata, and this is also the case in short-term experiments if the ozone concentration is beyond 0.1 $\mu l/l$. In contrast, an increased stomatal conductance could be observed after short-term exposure of plants with concentrations less than 0.1 $\mu l/l$ ozone (for a review see Saxe, 1991). These confusing data led Leonardi and Langebartels (1990) to suggest that ozone first promotes its own uptake at low concentrations, and, after longer exposure, induces the closure of the stomata. There is some speculation that automobile exhaust is also taken up mainly through stomata (Steenken, 1973).

Physiological effects of ozone

An interplay of several factors may be responsible for the generally observed reduction of stomatal conductance after ozone treatment. In the apoplastic space, ozone decomposes to superoxide radicals (O_2^-), hydrogen peroxide (H_2O_2) and hydroxyl radicals (HO^{-1}). As early as 1976 the ozone-induced lipid peroxidation was traced back to the action of such free radicals (Menzel, 1976).

Table 1. Effects of short-term ozone-fumigations on photosynthesis

Species	Concentration ($\mu l/l$)	Duration	Response (% control)	Reference
Avena sativa cv. "Titus"				Myhre et al (1988)
5-day-old plants	0.9	1h	96	
11-day-old plants	0.9	1h	92	
16-day-old plants	0.9	1h	92	
20-day-old plants	0.9	1h	88	
52/54-day-old plant	0.135-0.15	2h	75	
Nicotiana tabacum				Faensen and Thiebes (1983)
cv. "Bel W3"	0.25	1h	NS	
Phaseolus vulgaris				
cv. "Saxa"	0.15	2h	82	
Phaseolus vulgaris				Pell and Brennan (1973)
cv. "Pinto"	0.3	3h	78	
Avena sativa				Hill and Littlefield (1969)
	0.4	30 min	67	
	0.6	1h	64-12	
other species	0.4-0.7	0,5-2h	80-2	
Populus euamericana				Furukawa et al (1984a)
cv. "FS-51"	0.55	1h	72	
Populus euamericana				
cv. "Peace"	0.54	1h	88	
Helianthus annuus				
cv. "Russian Mammoth"	0.72	0,5-2h	57-42	
Vitis vinifera				Roper and Williams (1988)
cv. "Thompson seedless"	0.2-0.6	10h	100-45	
Pinus strobus	0.5-0.8	4h	93	Botkin et al (1972)

Consequently, ozone induces distinct changes in the chemical and physical properties of cellular membranes (Pauls and Thompson, 1980,1981). These in turn may generate drought stress which changes of the turgor pressure in the guard cells and lead to stomatal closure (Heath, 1980). One indication for such drought stress is the dramatic decrease in the ratio of fresh to dry weight that has been observed with increasing duration of ozone treatment (Sakaki et al, 1983).

In addition, drought stress triggers ethylene biosynthesis (for a review see Wang et al, 1990). Moreover, an increased ethylene content can also be observed after exposure to ozone (Elstner et al, 1985; Langebartels et al, 1991). Ethylene is known to reduce stomatal conductance (Taylor et al, 1988). Ozone treatment also increases the intercellular concentration of CO_2, that may also be involved in stomatal closure and the reduction of net photosynthesis (Reich, 1983; Reich et al, 1985, 1986; Temple, 1986; Omielan and Pell, 1988).

In fact, a reduced photosynthesis is a common feature in ozone-stressed plants (for a review see Darrall, 1989; Saxe, 1991; Table 1), as is a rapid loss of pigments (Sakaki et al, 1983). Additionally, the activity and content of ribulose-1,5-bisphosphate carboxylase/oxygenase (Rubisco) are both reduced by ozone (Pell and Pearson, 1983; Lehnherr et al, 1987; Pell et al, 1992). Dark respiration, especially after long-term exposure to ozone, is increased (Reich, 1983; Skärby et al, 1987; Amthor, 1988). Some authors speculate that this respiratory increase does not only represent a response to the destruction of the photosynthetic apparatus, but also part of a repair or defense mechanism in tissues damaged by ozone (Skärby et al, 1987; Amthor, 1988).

Actually, ozone not only triggers an increase in activity of the key enzymes of lignin biosynthesis, such as phenylalanine ammonium lyase (PAL) and coniferyl alcohol dehydrogenase (CAD; Sandermann et al, 1989; Heller et al, 1990), but also an accumulation of flavonoids or generally phytoalexins (Zielke and Sonnenbichler, 1990; Rosemann et al, 1991) as well as polyamines (Bors et al, 1989; Dohmen et al, 1990). Flavonoids and polyamines are considered as potent antioxidative compounds or radical scavengers (Bors et al, 1989; Dohmen et al, 1990). Ozone additionally increases vitamin E, vitamin C and glutathione (Mehlhorn et al, 1986), and activates glutathione reductase and ascorbate peroxidase, altogether indicators for a pronounced oxidative stress (Tanaka et al, 1988). The responses of other oxidative stress markers, superoxide dismutase (SOD) and catalase (CAT) are equivocal. For example, an increase in SOD activity has been documented in tobacco and spinach after a 4 hours treatment with 0.13 µl/l ozone (Decleire et al, 1984), whereas a 5 hours exposure of tobacco plants to 0.15 µl/l ozone obviously lead to a decrease of SOD activity (Kerner, 1990). Likewise, the results from long-term fumigation experiments with conifers are contradictory, too (Castillo et al, 1987; Polle and Rennenberg, 1991).

It is obvious that the effect of ozone on vegetation depends both on the time of exposure and concentration of the pollutant. However, ozone usually leads to stomatal closure, reduced net photosynthesis, an increased dark respiration and the activation of an antioxidative defense system.

Physiological effects of automobile exhaust and nitrogen oxides

Plants that were kept on open fields in close vicinity to busy highways reacted upon the pollutants with increased chlorophyll, sugar and free amino acid contents, an augmented ethylene production and peroxidase activity, a decreased ß-carotene, ascorbate and auxin concentration, and changes in fatty acid metabolism (Hoellwarth, 1981; Flueckinger et al, 1988). Automobile exhaust in concentrations of up to 1000 µl/l leads to a reduction in photosynthesis and impairment of stomata of *Picea abies* within only 15 minutes of treatment (Kammerbauer et al, 1986, 1987). The epistomal waxes had been structurally altered and clogged up the stomatal vestibles (Sauter et al, 1987). These toxic effects of the automobile exhaust were traced back to NO_x (Kammerbauer et al, 1987), that are solubilized in the water-saturated apoplastic space to form nitrous (HNO_2) and nitric acid (HNO_3). These compounds immediately dissociate into nitrate, nitrite and protons (Rowland et al, 1985), so that nitrate and nitrite accumulate in mesophyll cells of treated plants. This in turn induces the synthesis of nitrate (NR) and nitrite reductase (NiR) as well as glutamine synthetase (GS) and glutamate synthase (GOGAT; for a review see Wellburn, 1990).

Normally, the GS/GOGAT system assimilates most of the ammonium. However, plants stressed by NO_x can also metabolize ammonium via glutamate dehydrogenase (GDH; Wellburn, 1990). Relatively high endogenous nitrite concentrations (25mM) may reduce the activity of GS and GOGAT and lead to visible damage of plants (Yu et al, 1988). Moreover, nitrite represents a potent Hill oxidant and is able to release manganese from the water-evolving complex of thylakoid membranes, at least at higher concentrations (Wellburn, 1984).

The effect of NO_x on photosynthesis and dark respiration is highly dependent on exposure time and doses. As a rule, a short-term treatment with NO_x at concentrations beyond 0.3 µl/l increases chlorophyll content, dark respiration and net photosynthesis. Increasing concentrations of NO_x above 0.5 µl/l (or alternatively, lower doses applied for a longer time) affect these parameters negatively (for a review see Darrall, 1989; Wellburn, 1990; Saxe, 1991; table 2). NO has been defined as the most active component, being 20 times more toxic than NO_2 (Saxe, 1991). NO as a radical ($\cdot N=O$) can either accept or donate electrons. Its extreme toxicity may be explained on the basis of its reaction with ferridoxin. Indeed it could be proven that NO donates an electron to the reaction center of photosystem II (Petrouleas and Diner, 1990; Diner and Petrouleas, 1990). Additionally NO reacts with copper containing enzymes such as cytochrome c-oxidase (Brudvig et al, 1980). There is no increased formation of antioxidants by NO_x, in contrast to ozone and automobile exhaust. High concentrations of NO_x may , however, give rise to fatty acid peroxidation and there is evidence for an NO_x-dependent inhibition of lipid biosynthesis (Pryor and Lightsey, 1981; Malhotra and Khan, 1985).

The effect of nitrogen oxides on plants generally seems to be more dependent on exposure time and concentration of the pollutant than is the case with ozone. Though the data are far from being conclusive, we anticipate that nitrogen oxides can affect photosynthesis negatively and alter nitrogen fixation and dark respiration.

Molecular response to air pollutants

There is only scarce and circumstantial evidence for an influence of air pollutants on the structure or function of genes. For example, *Phaseolus vulgaris* reacts upon ozone with a depletion of chloroplast ribosomes (Chang, 1971, 1972), *Trifolium pratense* reduces its RNA, in particular 23 S rRNA content (Beckerson and Hofstra, 1979).

Table 2. Short term fumigations with nitrogen oxides and photosynthesis

Species	Concentration NO_2 ($\mu l/l$)	NO	Duration	Response (% control)	Reference
Lycopersicon esculentum		0.1	20h	89	Capron and Mansfield (1976)
cv. "Moneymaker"	0.1		20h	91	
	0.1	0.1	20h	82	
		0.5	20h	72	
	0.5		20h	68	
	0.5	0.5	20h	59	
Lycopersicon lycopersicon		1.0	3d	88	Bruggink et al (1988)
cv. "Abunda"		1.0	5d	68	
Phaseolus vulgaris,	0.5		2h	94	Srivastava et al (1975)
cv. "Pure Gold Wax"	1.0		2h	71	
	3.0		2h	53	
Lactuca sativa	0.5	2.0	30 min	90	Caporn (1989)
Helianthus annuus,	2.0		1h	90	Furukawa et al (1984b)
cv. "Russian Mammoth"		4.0	1h	80	
Medicago sativa,		2.0	2h	92	Hill and Bennett (1970)
cv. "Ranger"	2.0		2h	87	
	2.0	2.0	2h	75	
Avena sativa,		2.0	2h	93	
cv. "Park"	2.0		2h	88	
	2.0	2.0	2h	76	
Picea abies					Saxe and Murali (1989)
without pre-exposure		5.0	90 min	88	
with NO2 pre exposure		5.0	90 min	78	
without pre-exposure	5.0		90 min	75	
with NO pre-exposure	5.0		90 min	65	

The same effect has also been observed in leaves of *Betula alba* and *Cornus mas* that were exposed to pollutants generated by heavy traffic on nearby highways (Braun et al, 1991). Viroids have also been detected in pollutant-stressed plants (Schuhmacher et al, 1984; Beuther et al, 1988; Koester et al, 1988). To our knowledge, there is only one paper which reported an increased expression of chitinase, extensin and ß-1,3-glucanase genes in plants after ozone treatment (Langebartels et al, 1991).

Planning and evaluation of air pollutant experiments

This short introduction illustrates the present situation of pollutant research in plants. Though lots of data on the effects of air pollutants on plants are available, we are far from understanding the precise mechanism(s) of how pollutants affect plants. However, the present data allow the conclusion that both planning and evaluation of experiments with pollutants may cause some problems.

First of all, long-term pollutant stress has to be neatly discriminated from short-term, acute stress, since the plant's responses to both influences are qualitatively different (Ehlinger, 1989). On one hand, long-term fumigations normally serve to test the ability of a plant to survive under the influence of pollutants and do not add to our understanding of the reaction mechanism of these compounds. Short-term fumigations, on the other, allow to decipher reaction mechanisms of pollutants down to the molecular detail. One of the problems of previous physiological experiments becomes apparent in table 1 and 2. In order to induce recognizable effects in plants, short-term treatments frequently employ extreme concentrations of pollutants that also lead to visible damage. It is obvious that such experiments and any conclusions drawn from the results do not at all reflect the situation of a plant exposed to *realistic* pollutant stress.

Second, the technique of fumigation is of outstanding importance for the evaluation of experiments. Thus, open field experiments hardly allow to trace symptoms back to specific compounds, since too many of the physiologically effective parameters vary too widely. This also holds for "open-top" chambers that allow a more or less defined pollutant stress to be applied, but do not keep all physiologically relevant factors constant. In our view, only precisely controlled conditions, such as prevail in climate chambers, really justify to draw firm conclusions about the reaction mechanisms of particular compounds. This is especially true for complex air pollutant mixtures that may vary in composition depending on the type of engine and driving conditions (Graedel, 1979). Moreover, the design of fumigation experiments is important. For example, it is rather pointless to exercise the commonly used concurrent exposure of plants with NO_x and ozone, since ozone can and will react with NO to NO_2 and $1/2\ O_2$ in light (Rowland et al, 1985).

Photooxidant stress and gene expression

It is apparent that practically nothing is known about the effects of air pollutants on the structure and function of plant genes. We are also ignorant about the role of secondary stressors such as wounding of plants or their attack by pathogens. Therefore we started a series of experiments aiming at characterizing the effect(s) of photooxidants on plant genes, both under precisely controlled conditions and with realistic concentrations of pollutants. Using Southern-, Northern- and DNA fingerprint analyses the problem of whether plants will be influenced genetically by automobile exhaust directly or also through secondary products has been addressed. We selected three groups of genes that are essential for plant life:

1. nuclear photosynthesis genes, that have a key function in energy metabolism of plants.
2. nuclear wound response and defense genes, that control the sequence of events leading to wound healing and encode proteins engaged in defense reactions against plant pathogens (e.g. fungi).
3. nuclear antioxidative defense genes whose products detoxify oxidative compounds.

Experimental conditions

For the detection of changes in the structure or methylation status of a gene, DNA was isolated from tobacco plants (*Nicotiana tabacum* cv. SR-1) that were left untreated (control), or treated with automobile exhaust and/or ozone. The DNA was restricted with methyl-sensitive restriction endonucleases *Hpa*II and *Msp*I, the resulting fragments separated by agarose gel electrophoresis, blotted onto Nylon membranes and hybridized to the probes listed in Table 3.

Additionally DNA fingerprint techniques were applied to screen for pollutant-induced changes in highly repetitive DNA-sequences (e.g. microsatellites). Genomic DNA was restricted with *Hin*fI and the resulting fragments were separated by agarose gel electrophoresis, blotted onto Nylon membranes and hybridized to the radiolabeled synthetic oligodeoxynucleotides $(GATA)_4$ and $(CA)_8$.

Northern blot hybridization was used to detect any changes in the steady-state levels of mRNA from the different genes in *Nicotiana tabacum* cv. SR-1 plants, that were exposed to automobile exhaust and/or ozone. Since air pollutants may well increase the sensitivity of plants towards pathogens or insect attack (Krupa and Manning, 1988), we also determined the expression of genes in plants that were additionally exposed to a secondary stress during

Table 3. DNA probes used in air pollutant experiments

DNA Probe	Type	Enzyme	Cloning procedures or sequences described by
rbcs	cDNA	small subunit of ribulose-1,5-bisphosphate carboxylase/Oxygenase	Cashmore (1983)
cab	cDNA	light-harvesting chlorophyll a/b-binding protein	Coruzzi et al (1983)
ST-LS1	Genomic	10 kd-protein of water-evolving complex of PS II	Eckes et al (1986)
CHN 50	cDNA	chitinase	Shinshi et al (1987)
GL 43	cDNA	ß-1,3-glucanase	Shinshi et al (1988)
PAL	cDNA	phenylalanine-ammonium-lyase	Somssich et al (1986)
CHS	cDNA	chalcone-synthetase	Somssich et al (1986)
Sod 1	cDNA	mitochondrial Mn-superoxide-dismutase	Bowler et al (1989)
Sod 2	cDNA	chloroplast Fe-superoxide-dismutase	Van Camp et al (1990)
Sod 3	cDNA	cytosolic Cu/Zn-superoxide-dismutase	Tsang et al (1991)

treatment with automobile exhaust. Secondary stress was imposed by mechanical wounding or application of salicylic acid that simulate pathogen attack, since it functions as endogenous elicitor (Malamy et al, 1990; Yalpani et al, 1991; Ward et al, 1991).

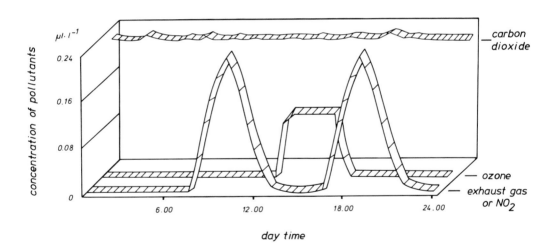

Figure 1. Treatment of Nicotiana tabacum *cv. SR-1 with air pollutants.*
The plants were exposed to a two-peak fumigation regime with a maximum concentration of 0.24 µl/l NO_x (exhaust gas) or nitrogen dioxide (NO_2), and an intermittent ozone concentration of 0.12 µl/l that lasted four hours per day. CO_2 was applied continuously at a concentration of 300 µl/l a day (according to Schaub et al, 1991).

Realistic short-term fumigation regimes

The daily course of pollutant changes characteristic of urbane regions was simulated in a special regime for fumigation that was designed by Schaub et al (1991, Fig.1). Cloned tobacco plants (*Nicotiana tabacum* cv. SR-1) were grown up in a green-house for six weeks, then transferred to exposure chambers and adapted to the new environment by treatment with clean air for 10 days. Subsequently the plants were exposed to air pollutants as detailed in Figure 1.

The chambers were illuminated for 10 h by five 1000 W mercury-halide vapor lamps (Osram HQI-T 1000/D; quantum flux density: 650 µE/m sec, PAR). Day/night temperature (22/18°C) and relative humidity (65%) were identical for all chambers. Controls were supplied with filtered air. The actual concentration of CO_2, NO_x and O_3 was continuously measured and regulated as described by Schaub et al (1990). The automobile exhaust was produced under controlled conditions in a dynamical engine-test stand using the FTP-75 drive-cycle (EPA, 1981; Schaub et al, 1991). After 48 h of storage the diluted gas contained 0.064 ml/l hydrocarbons (HC), 0.15 ml/l NO_x (0.05 ml/l NO, 0.1 ml/l NO_2), 0.8 ml/l CO, 1% CO_2 and 19.3% O_2. The NO_x/HC ratio was 2.3.

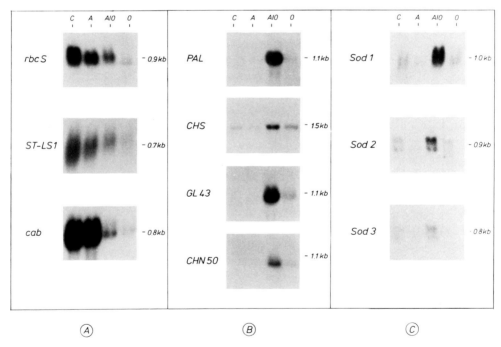

Figure 2. Northern blot analysis of photosynthesis (A), pathogen defense (B) and antioxidative defense genes after air pollutant treatment. Total RNA was isolated from control (C) and polluted plants (A: automobile exhaust; A/O: exhaust gas/ozone; O: ozone), electrophoresed and blotted onto Nylon membranes. The RNA was hybridized to the ^{32}P-labeled probes described in Table 3. Based on analysis of intact plants.

Transcript levels of photosynthesis, pathogen defense and antioxidative defense genes

Though the plants did not exhibit visible injuries at the end of the 48 hours fumigation period, Northern blot analyses detected considerable changes in the steady-state mRNA levels of the different genes (Fig. 2-4). However, we were not able to detect any structural changes or methylation modifications in DNA (data not shown).

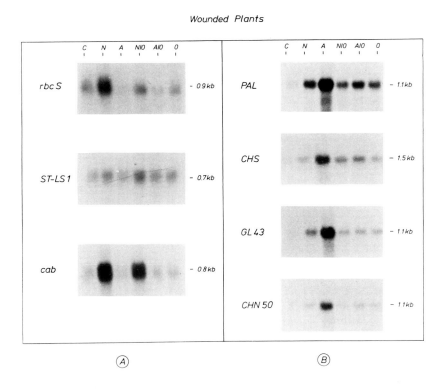

Figure 3. Northern blot analysis of photosynthesis (A) and pathogen defense genes (B) after air pollutant treatment and wounding .

Plants were wounded by treating their leaves with a razor blade one hour before the fumigation experiments. For elicitor treatment the plants were wounded as described, and then sprayed with salicylic acid (0.5 mM). The liquid was evenly distributed on the leaf surface by brief rubbing. Then total RNA was isolated from control (C) and polluted plants (N: NO_2; A: automobile exhaust; N/O: NO_2/O_3; A/O: exhaust gas/ozone; O: ozone), electrophoresed and blotted onto Nylon membranes. The RNA was hybridized to ^{32}P-labeled probes described in Table 3.

Elicitor-treated Plants

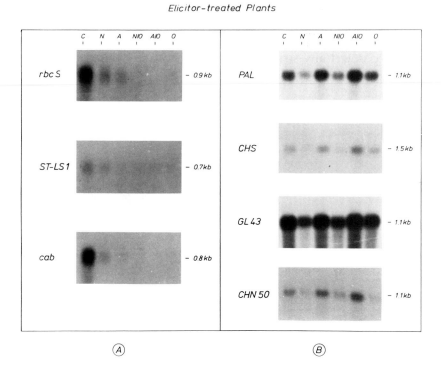

Figure 4: Northern blot analysis of photosynthesis (A) and pathogen defense genes (B) after elicitor trweatment. Legend as for Figure 3.

Densitometric scanning of autoradiographs was used to quantify the steady-state levels of mRNAs from the different genes. Potential differences in the amounts of total RNA applied to each lane were corrected by densitometry of the ethidium bromide-stained agarose gels and the autoradiographs. The results of these experiments illustrate that:

1. Air pollutants reduce the expression of the three photosynthesis genes (*rbcs, cab, ST-LS1* gene), with automobile exhaust being less effective (reduction about 40% of control) than ozone (reduction about 90%). See figure 5 A.
2. Air pollutants, especially automobile exhaust in combination with ozone, induce wound response and defense genes (CHS, PAL, chitinase and ß-1,3-glucanase) to high expression levels (Fig. 5 B). Exhaust gas alone elicits only the gene coding for ß-1,3-glucanase (to about 160%), whereas the expression of the CHS gene is reduced to some 35%.
3. The expression of all three superoxide dismutase genes is induced by

ozone, especially in combination with gasoline exhaust (Sod 1: up to 760%, Sod 2: 143%). If exposed to automobile exhaust alone the plants exhibit reduced superoxide dismutase mRNA steady-state levels (Sod 1: 70%, Sod 3: 30%, see Fig. 5 C).
4. Moreover, additional wound or pathogen stress reduces the expression of photosynthesis genes (rbcs, cab, ST-LS1) to a greater extent than exhaust gas and/or ozone stress alone (Fig. 6 A,7 B).

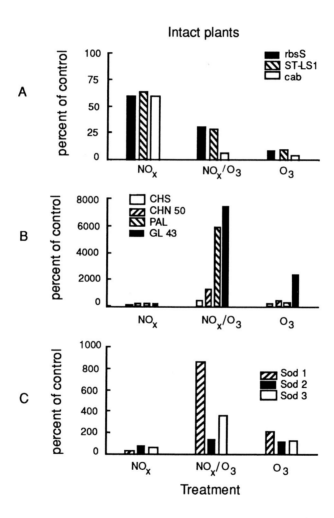

Figure 5. Transcript levels of photosynthesis (A), pathogen defense (B) and antioxidative defense genes (C) after 48h fumigation with air pollutants.
Tobacco SR-1 plants were treated with exhaust gas, singly (NO_x) or in combination with ozone (NO_x/O_3) and ozone (O_3) as described in the legend to figure 1. For each gene the steady-state mRNA level was plotted against the corresponding control (100%).

5. A simultaneous treatment of plants with exhaust gas (alone or in combination with ozone) and wounding or elicitor increases the steady-state mRNA levels of all four wound response and defense genes (coding for CHS, PAL, chitinase and ß-1,3-glucanase). This also holds for ozone treatment in combination with wounding or elicitor application (Figs. 6B; 7B).

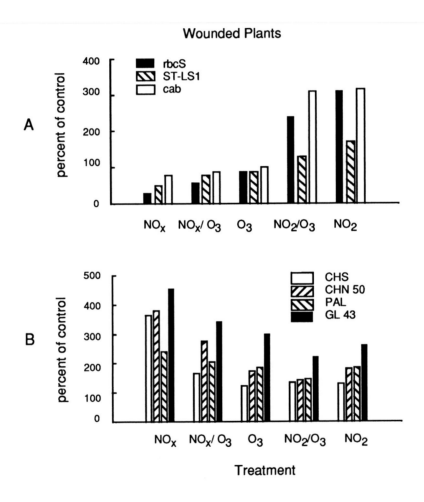

Figure 6. Transcript levels of photosynthesis (A) and pathogen defense genes (B) after 48h of wounding and air pollutant treatment.
Tobacco SR-1 plants were fumigated with various combinations of air pollutants as outlined in the legend of figure 1. For each gene the steady-state mRNA level was plotted against a corresponding wounded control (100%). NO_x: exhaust gas; NO_x/O_3: exhaust gas/ozone; O_3: ozone; NO_2: nitrogen dioxide; NO_2/O_3: combination of nitrogen dioxide and ozone.

6. NO$_2$ affects plants differently from automobile exhaust. Thus, NO$_2$ alone or in combination with ozone, if applied to wounded plants enhances the expression of all three photosynthesis genes (Fig. 6 A). Wound response and defense genes are induced but not expressed to the high levels in plants exposed to gasoline exhaust (with or without ozone; see Fig. 6 B).

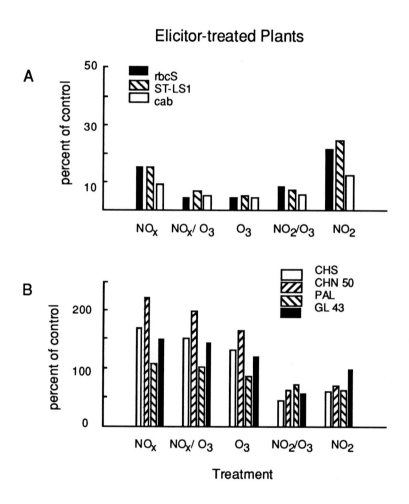

Elicitor-treated Plants

Figure 7. Transcript levels of photosynthesis (A) and pathogen defense genes (B) after 48h of elicitor (0.5 mM salicylic acid) and air pollutant treatment.
Tobacco SR-1 plants were exposed to automobile exhaust, singly (NO$_x$) or in combination with ozone (NO$_x$/O$_3$), ozone (O$_3$), nitrogen dioxide (NO$_2$) and a combination of nitrogen dioxide and ozone (NO$_2$/O$_3$) as described in the legend to Figure 1. For each gene the steady-state mRNA level was plotted against a corresponding, salicylic acid-treated control (100%).

A combination of NO$_2$ and elicitor treatment tends to diminish the effect of automobile exhaust such that e.g. the steady-state levels of the photosynthesis genes are less affected (Fig. 7 A). As compared to exhaust gas stress, alone or in combination with ozone, the above treatment reduced the expression of the defense genes (Fig. 7 B).

Discussion

Stress induces the expression of groups of genes

For a coordinated regulation of metabolic processes in a cell it is essential to express functional groups of genes coordinately. In situations of stress, such as stress imposed by air pollutants, functionally related genes are in fact simultaneously induced. For example, photosynthesis genes are repressed, pathogen defense genes are induced and antioxidative defense genes are transcribed at higher rates, if plants are exposed to ozone (Fig. 5 A-C).

Similar responses of plants were also found with other stressors. Thus, pathogen attack represses the synthesis of the small and large subunit of Rubisco and a 32 kd thylakoid protein in *Hordeum vulgare* (Higgins et al, 1985), and UV-B irradiation stress lowers the transcript contents of *rbcs* and *cab* genes in *Pisum sativum* (Jordan et al, 1991, 1992). Defense genes encoding chitinase and ß-1,3-glucanase are also increasingly expressed in response to elicitors generated after pathogen attack (for a review see Dixon, 1986), and ozone (Langebartels et al, 1991).

Ethylene is also a potent inductor of gene expression (Collinge and Slusarenko, 1987; Boller, 1987), and was shown to be active in the expression of wound response genes (e.g. those coding for phenylalanine ammonium lyase, PAL, and chalcone synthase, CHS; Ecker and Davis, 1987). PAL is a key enzyme in plant wound reactions and its synthesis is induced by elicitors coordinately with CHS (for a review see Dixon, 1986). Both PAL and CHS genes are activated by visible or UV light (Smith, 1990). Another group of genes, encoding superoxide dismutases of mitochondrial and cytosolic location, is induced by pathogen stress in *Nicotiana plumbaginifolia* (Bowler et al, 1989). These few examples document, that plants usually react upon stressors of whatever kind with the activation or silencing of groups of genes. The coordinate expression of physically separate, but functionally related genes in response to very different external stimuli suggests a common molecular mechanism.

Photooxidant stress induces an antioxidative defense system

There is evidence that the primary reaction of a plant upon air pollutants, such as automobile exhaust or ozone, is similar to wounding or pathogen attack. One common effect of all these stressors is the disintegration of cellular membranes. The mechanism of this deleterious reaction is known in some detail. For example, ozone decomposes into superoxide radicals (O_2^-), hydrogen peroxide (H_2O_2), and the extremely reactive hydroxyl radical ($OH\cdot$), once it becomes solubilized in the apoplastic space (Hoigne and Bader, 1975; Heath, 1979; Grimes et al, 1983). Moreover, the hydrocarbons of exhaust gas contain a series of compounds that undergo different oxidation processes within the plant (Durmishidze et al, 1985) or disintegrate to radicals, as e.g. formaldehyde (Van Haut et al, 1979). The accumulation of such radicals and activated oxygen species entails a peroxidation of lipids and a subsequent loss of selective membrane permeability which finally leads to disturbances of compartmentalization and the change in properties of membrane-bound enzymes such as plasma membrane ATPase (Pauls and Thompson, 1980, 1981; Dominy and Heath, 1985). Changes in membrane permeability and activity of plasma membrane ATPase are also induced by elicitors that simulate a pathogen attack (Sze, 1985; Low and Heinstein, 1986).

Both air pollutants and elicitors have in common that they lead to the accumulation of oxyradicals in the extracellular space, since these radicals cannot cross the plasma membrane, with the exception of H_2O_2. Therefore, the elimination of these potentially dangerous oxyradicals would primarily depend on extracellular mechanisms. In fact, potential radical scavengers such as vitamin E, ascorbate and glutathione accumulate in the apoplastic space after ozone stress (Tanaka et al, 1985; Mehlhorn et al, 1986). Ozone additionally enhances the formation of callose, lignin, extensin, polyphenols, phytoalexins and polyamines (Rosemann et al, 1991; Heller et al, 1090; Bors et al, 1990). Some of these compounds act as scavengers of oxyradicals (Bors et al, 1990), either alone or in conjugates with e.g. hydroxy cinnamic acid and polyamines.

Our present results support these data. Ozone, and especially a mixture of ozone and car exhaust, induce wound response and defense genes or lead to their enhanced expression (Fig. 5 B,C). We regard the expression of PAL, CHS and ß-1.3-glucanase as a consequence of an increased oxidative stress in the apoplastic space. Nevertheless, it should be kept in mind that lipid peroxidation with subsequent damage of membranes also build up an oxidative stress within the cell. As a consequence, we find an increased expression of the genes encoding the cytosolic SOD (Fig. 5C). Again, the combined action of ozone and car exhaust is by far more effective than both components alone. We observe a slight but distinct decline of the expression of SOD and CHS genes under ozone or exhaust gas stress alone, probably a consequence of a diminished oxidative stress.

It is obvious that the plant reacts upon the generation of oxyradicals by whatever environmental stress, e.g. air pollutant stress, with an increased synthesis of compounds with radical scavenger properties.

Photooxidant stress increases mitochondrial respiration

The massive increase in metabolic activity of plants stressed by air pollutants requires adequate energy supply. One of the energy sources, photosynthesis, cannot be exploited since e.g. ozone reduces both activity and content of ribulose-1,5-bisphosphate carboxylase/oxygenase with subsequent decrease in net photosynthesis (Saxe, 1991) and concomitant increase in dark respiration (Reich, 1983; Skärby et al, 1987; Amthor, 1988). Dark respiration is therefore considered as the major energy source in plants stressed by air pollutants. One indicator for an increased respiration, the increased cytochrome c-oxidase activity, correlates positively with the activity and expression of a mitochondrial SOD in *Nicotiana plumbaginifolia* (Bowler et al, 1989). The expression of mitochondrial SOD in tobacco stressed with ozone and/or automobile exhaust (Fig. 5C) suggests, that its induction is more likely a consequence of increased respiration rather than an effect of pollutant- related oxyradicals in mitochondria.

The primary effects of photooxidants are extra-chloroplastic

Up to date, the decrease in quantity and activity of Rubisco has been explained by its increased proteolysis as a consequence of oxygen stress (Pell et al, 1992). However, our results document that the expression of the *rbcs* genes, or in other words the synthesis of Rubisco are affected. The transcription of *rbcs, cab* and *ST-LS1* genes presupposes an intact chloroplast (Simpson et al, 1986; Fluhr et al, 1986; Stockhaus et al, 1987) and regulation by light via the phytochrome system and other photoreceptors (e.g. cryptochrome, UV-B photoreceptor; Tobin, 1987; Mohr, 1987). For example, the photooxidative destruction of pigments by light in carotinoid-free mustard chloroplasts entails a decreased transcription of rbcs and cab genes (Oelmueller et al, 1986). There is ample evidence that chlorophyll degradation can only proceed if the thylakoid membranes are substantially damaged (Sakaki et al, 1983). Additionally, it has been reported that a distinct pigment loss occurs only after 5 days of ozone stress. The authors therefore suggested that the primary effect of ozone is extra-chloroplastic (Price et al, 1990).

We therefore do not interpret the reduced transcript levels of the photosynthesis genes in our experiments (Fig. 5A) as a consequence of an ozone-induced disintegration of thylakoid membranes and a damage of the chloroplasts. Instead, we propose that an alteration or modification of phytochrome or other photoreceptors, caused by the oxidative reactions in the plasmalemma and cytosol, is the primary effect of photooxidants. As a result, photosynthesis genes are repressed and photosynthesis is subsequently reduced even if the chloroplasts are not damaged. It is remarkable that genes encoding chloroplastic SOD remain unaffected even after 48 hours of air pollutant stress (Fig. 5C), in contrast to both cytoplasmic and mitochondrial SOD enzymes, which are highly expressed.

Air pollutants and secondary stress

A striking symptom of the recent forest decline is the increase in number of trees attacked by pathogens, e.g. fungi. We hypothesize that air pollutants, especially ozone, may increase a plant's susceptibility towards pathogens (see also Krupa and Manning, 1989). Our results indeed support these data. Thus, intact plants react more strongly to ozone than to automobile exhaust stress (Fig. 5). If such plants are additionally exposed to a secondary stress, e.g. wounding, then their response toward exhaust gas is much stronger (Fig. 6). Secondary stress then seems to enhance air pollutant stress in plants. For example, wounding obviously removes barriers that protect the plant from the action of air pollutants. These compounds, in particular ozone, are taken up through stomata. However, car exhaust components in high concentrations may lead to stomatal closure (Sauter et al, 1987). One of this compounds, NO_x does not influence stomata regulation (Darrall, 1989; Saxe, 1991), while ozone promotes its own uptake via a positive feed-back mechanism (Langebartels et al, 1991). As soon as wounding destroyed protective structures such as the cuticle, the uptake of air pollutants is no more dependent on stomatal conductance. This obviously facilitates the dissolution of exhaust components in the apoplastic space with their subsequent decay into radicals and other oxidative substances. This in turn leads to an accumulation of oxyradicals and other reactive compounds within the apoplastic space and cells, and to a dramatic suppression of photosynthesis genes after exhaust gas (and/or ozone) treatment and wounding (Fig. 6A). This effect is specific, since the identical treatment drastically increased the transcript levels of different wound response and defense genes (Fig. 6B). Simulation of pathogen attack by salicylic acid treatment superimposed on air pollutant stress (ozone and/or car exhaust) also leads to similar effects (Fig. 7B).

Obviously, the combination of air pollutant and secondary stress, e.g. wounding or pathogen attack, exerts a stronger effect on plant gene expression than treatment with either one of the stressors. This may explain, why the sensitivity of a plant to pathogens is increased if it is stressed by air pollutants, especially ozone (Krupa and Manning, 1989).

Hydrocarbons - the truly toxic components of exhaust gas?

Emissions from automobiles have frequently been connected with the recent forest decline. To determine the influence of hydrocarbon compounds in automobile exhaust on gene expression, some of the (wounded and salicylic acid-treated) plants in the present study were fumigated alternatively with NO_2 or exhaust gas (both, singly or in combination with ozone). The main difference between a single application of NO_2 and automobile exhaust fumigation is the absence of the hydrocarbon compound in the former, especially since CO, another main component of exhaust gas, has no negative effect on plants, since it is oxidized to CO_2 (Ducet and Rosenberg, 1962) or metabolized to serine (Chappelle and Krall, 1961).

Our results show that the four wound response and defense genes are altogether expressed at a higher rate, if the corresponding plant has been exposed to automobile exhaust. NO_2 on the contrary had no effect (Fig 6B). Again, exhaust gas stress reduces photosynthesis gene activity, NO_2 alone increases it (Fig. 6A). Fumigation of plants with NO_2 (and/or ozone) and simultaneous treatment with an elicitor depress the transcript levels of photosynthesis and defense genes, but less pronounced as compared to the extreme low levels characteristic for plants treated with exhaust gas (Fig. 7 A,B). We therefore suggest that hydrocarbons are favorite candidates for the phytotoxic effect of exhaust gas. The hydrocarbons arise from incomplete combustion of gasoline, and even a catalytic converter can remove only part of it. Thus in future, the design of engines should consider the reduction of hydrocarbons as a necessary element, since specific components of the HC fraction not only influence vegetation (see above), but are also toxic and carcinogenic in animal and human cells (Alsberg et al, 1984).

Perspectives

The present results document that air pollutants do not only interfere with morphological or physiological processes of plants but also influence gene expression at low concentrations and in a relatively short time. Genes encoding proteins of the photosynthetic apparatus are silenced, whereas wound response and defense genes are activated. This differential effect of air pollutants on plant defense genes poses a series of questions for future research:

1. How do other groups of functionally related genes behave under realistic exhaust gas stress?

2. What is the mechanism of action of the active components of automobile exhaust, the hydrocarbons and - as secondary product - ozone? Do they interact with transcription factors (Weising and Kahl, 1991), or with the cascade of signal transduction from membranes to genes?

3. What is the role of cellular antioxidants in detail, and what is the precise function of superoxide dismutases, catalases and peroxidases in pollutant stressed plants?

4. Is only transcription of genes influenced by air pollutants, or are the genes themselves mutated or simply modified? Our experiments show that short-term fumigations with exhaust gas or components of it do not cause

any detectable changes in DNA structure (e.g. methylation of cytosine residues within specific sequences). However, long-term exposure to automobile exhaust might be different, since it induces premature senescence.

5. Last not least, we propose to perform experiments with well-defined, i.e. cloned model plants under precisely controlled conditions and with realistic concentrations of air pollutants in order to collect reliable and really meaningful data about the molecular reactions of plants towards pollutants.

Acknowledgments

Research of the authors was supported by Bundesministerium für Forschung und Technologie (BMFT grant FKZ 0339190F), Bonn, Germany. We appreciate the cooperation of Professor H. Schaub of this Department who developed the exposure chambers.

References

Alsberg, T., Stenber, U., Westerholm, R. , Strandell, M., Rannung, U., Sundvall, A., Romert, L., Bernson, V., Pettersson, B., Toftgard, R., Franzen, B., Jansson, M., Gustafsson, J.A., Egebäck, K.E. & Tejie, G. (1984) *Environmental Science & Technology* **19**, 43-50.

Amthor, J.S. (1988) *New Phytologist* **110**, 319-325.

Beckerson, D.W. & Hofstra, G. (1979) *Canadian Journal of Botany* **57**, 1940-1945.

Bennett, J.H. & Hill, A.C. (1973) *Journal of Environmental Quality* **2**, 256-530.

Beuther, E., Köster, S., Loss, P., Schumacher, J. & Riesner, D. (1988) *Journal of Phytopathology* **121**, 289-302.

Boller, T. (1987) *Oxford Surveys of Plant Molecular & Cell Biology* **5**, 145-174.

Botkin, D.B., Smith, W.H. & Carlson, R.W. (1972) *Journal of the Air Pollution Control Association* **21**, 778-780.

Bowler, C., Alliotte, T., DeLoose, M., Van Montagu, M. & Inze, D. (1989) *EMBO Journal* **8** ,31-38.

Braun, S, Flueckinger, W. & Oertli, J.J. (1981) *Mitteilungen der deutschen Gesellschaft für allgemeine angewandte Entomologie* **3**, 138-139.

Brudvig, G.W., Stevens, T.H. & Chan, S.I. (1980) *Biochemistry* **19**, 5275-5285.

Bruggink, G.T., Wolting, H.G., Dassen J.H.A. & Bus, V.G.M. (1988) *New Phytologist* **110**,185-191.

Caporn, S.J.M. (1989) *New Phytologist* **111**, 473-481.

Cashmore, A.R. (1983) *Cell* **17**, 383-388.

Castillo, F.J. & Greppin, H. (1988) *Environmental & Experimental Botany* **28**, 231-238.

Chang, C.W. (1971) *Phytochemistry* **10**, 2863-2868.

Chang, C.W. (1972) *Phytochemistry* **11**, 1347-1350.

Chappelle, E.W. & Krall, A.R. (1961) *Biochemica Biophysica Acta* **49**, 578-580.

Collinge, D.B. & Slusarenko, A.J. (1987) *Plant Molecular Biology* **9**, 389-410.

Darrall, N.M. (1989) *Plant, Cell & Environment* **12**, 1-30.

Decleire, M., De Cat, W., De Temmerman, L. & Baeten, H. (1984) *Journal of Plant Physiology* **116**, 147-152.

Diner, B.A. & Petrouleas, V. (1990) *Biochimica et Biophysica Acta* **1015**, 141-149.

Dixon, R.A. (1986) *Biological Reviews* **61**, 239-291.

Dohmen, G.P., Koppers, A. & Langebartels, C. (1990) *Environmental Pollution* **64**, 375-383.

Dominy, P.J. & Heath, R.L. (1985) *Plant Physiology* **77**, 43-45.

Ducet, G. & Rosenberg, A.I. (1962) *Annual Review of Plant Physiology* **13**, 171-200.

Durmishidze, S.V., Chrikishvili, D.I., Beriashvili, T.V., Maisuradze, Ts.M. & Gugunishvili, G.Sh. (1985) *Prikl. Biokhim. Mikrobiol.* **21** , 318-323.

EPA (1981) Environmental Protection Agency (EPA). *Federal Code of Regulation* part 85. U.S. Government Printing Office, Washington.

Ecker, J.R. & Davis, R.W. (1987) *Proc. Natl. Acad. Sci. (USA)* **84**, 5202-5206.

Eckes, P., Rosahl, S., Schell, J. & Willmitzer, L. (1986) *Molecular & Gen. Genetics* **205**, 14-22.

Ehlinger, J.R. (1989) In: *Ecological Genetics and Air Pollution*. Taylor, G.E. (ed.). pp 203-208. Springer-Verlag New York, Heidelberg

Elstner, E.F., Osswald, W. & Youngman, R.J. (1985) *Experientia* **41**, 591-597.

Faensen-Thiebes, A. (1983) *Angewandte Botanik* **57**, 181-191.

Fluhr, R., Kuhlemeier, C., Nagy, F. & Chua N.H. (1986) *Science* **232**, 1106-1112.

Flueckinger, W., Braun, S. & Bolsinger, M. (1988) In: *Air Pollution and Plant Metabolism*. Schulte-Hostede, S., Darrall, N.M., Blank, L.W. & Wellburn, A.R. (ed.). pp 366-380. Elsevier Applied Science, London.

Furukawa, A., Katase, M., Ushijima, T. & Totsuka, T. (1984a) *Research Report from the National Institute for Environmental Studies* **65**, Ibaraki, Japan pp 77-87.

Furukawa, A., Yokoyama, M., Ushijima, T. & Totsuka, T.(1984b) *Research Report from the*

National Institute for Environmental Studies **65**, Ibaraki, Japan pp 89-98.

Graedel, T.E. (1979) *Reviews of Geophysical Space Physics* **17**, 937-947.

Grimes, H.D., Perkins, K.K. & Boss, W.F. (1983) *Plant Physiology* **72**, 1016-1020.

Heath, R.L. (1979) *Toxicological Letters* **4**, 449-453.

Heath, R.L. (1980) *Annual Review of Plant Physiology* **31**, 395-431.

Heller, W., Rosemann, D., Osswald, W.F., Benz, B., Schönwitz, R., Lohwasser, K., Kloos, M. & Sandermann, H. (1990) *Environmental Pollution* **64**, 353-366.

Higgins, M.C., Manners, J.M. & Scott, K.J. (1985) *Plant Physiology* **78**, 891-894.

Hill, A.C. & Bennett, J.H. (1970) *Atmospheric Environment* **4**, 341-348.

Hill, A.C. & Littlefield, N. (1969) *Environmental Science & Technology* **3**, 52-56.

Hoigne, J. & Bader, H. (1975) *Science* **190**, 782-784.

Jordan, B.R., Chow, W.S., Strid, A. & Anderson, J.M. (1991) *FEBS Letters* **284**, 5-8.

Jordan, B.R., He, J., Chow, W.S. & Anderson, J.M. (1992) *Plant, Cell & Environment* **15**, 91-98.

Kammerbauer, H., Selinger, H., Römmelt, R., Ziegler-Jöns, A., Knoppik, D. & Hock, B. (1986) *Environmental Pollution* **42**, 133-142.

Kammerbauer, H., Selinger, H., Römmelt, R., Ziegler-Jöns, A., Knoppik, D. & Hock, B. (1987) *Environmental Pollution* **48**, 235-243.

Kerner, K. (1990) Biochemische und physiologische Untersuchungen zur unterschiedlichen Ozontoleranz der Tabaksorten (*Nicotiana tabacum* L.) Bel B und Bel W3, *Thesis*, Universität München, Germany.

Koziol, M.J., Whatley, F.R. & Shelvey, J.D. (1988) In: *Air Pollution and Plant Metabolism.* Schulte-Hostede, S., Darrall, N.M., Blank, L.W. & Wellburn, A.R. (ed.). pp 148-168. Elsevier Applied Science, London.

Krupa, S.V. Manning, W.J. (1988) *Environmental Pollution* **50**, 101-137.

Köster, S., Beuther, E. & Riesner, D. (1988) *Journal of Phytopathology* **121**, 303-312.

Langebartels, C., Kerner, K., Leonardi, S., Schraudner, M., Trost, M., Heller, W. & Sandermann, H. Jr. (1991) *Plant Physiology* **95**, 882-889.

Lehnherr, B., Grandjean, A., Machler, F. & Fuhrer, J. (1987) *Journal of Plant Physiology* **130**, 189-200.

Leonardi, S. & Langebartels, C. (1990) *Water, Air & Soil Pollution* **54**, 143-153.

Low, P.S. & Heinstein, P.F. (1986) *Acta Biophysica et Biochemica* **249**, 472-479

Malamy, J. Carr, J.P., Klessig, D.F. & Raskin, I. (1990) *Science* **250**, 1002-1004.

Malhotra, S.S. & Khan, A.A. (1984) In: *Air Pollution and Plant Life.* Treshow, M. (ed.), John Wiley & Sons, New York pp 113-157.

Mehlhorn, H., Seufert, G., Schmidt, A. & Kunert, K.J. (1986) *Plant Physiology* 82, 336-338.

Menzel, D.B. (1976) In: *Free Radicals in Biology.* VOL. II, Pryor, W.A.(ed.), Academic Press, London pp 181-202.

Mohr, H. (1987) In: *Phytochrome and Photoregulation in Plants.* Furuya, M. (ed.), Academic Press, Tokyo, Orlando, London pp 333-348.

Myhre, A., Forberg, E., Aarnes, H. & Nielsen, S. (1988) *Environmental Pollution* 53, 265-271.

Oelmueller, R. & Mohr, H. (1986) *Planta* 167, 106-113.

Omielan, J.A. & Pell, E.J. (1988) *Canadian Journal of Botany* 66, 745-749.

Pauls, K.P. & Thompson, J.E. (1980) *Nature* 283, 504-506.

Pauls, K.P. & Thompson, J.E. (1981) *Physiologia Plantarum* 53, 255-262.

Pell, E.J. & Brennan, E. (1973) *Plant Physiology* 51, 378-381.

Pell, E.J. & Pearson, N.S. (1983) *Plant Physiology* 73, 185-187.

Pell, E.J., Eckardt, N. & Enyedi, A.J. (1992) *New Phytologist* 120, 397-405.

Petrouleas, V. & Diner, B.A. (1990) *Biochimica et Biophysica Acta* 1015, 131-140.

Polle, A. & Rennenberg, H. (1991) *New Phytologist* 117, 335-343.

Price, A., Young, A., Beckett, P., Britton, G. & Lea, P. (1990) In: *Current Research in Photosynthesis IV*, Baltscheffsky, M. (ed.). pp 595-598. Kluwer Academic Publ., Dordrecht, Boston, London.

Pryor, W.A. & Lightsey, J.W. (1981) *Science* 214, 435-437.

Reich, P.B. (1983) *Plant Physiology* 73, 291-296.

Reich, P.B. (1987) *Tree Physiology* 3, 63-91.

Reich, P.B. & Amundson, R.G. (1985) *Science* 230, 566-570.

Reich, P.B., Schoettle, A.W., Raba, R. & Amundson, R.G. (1986) *Journal of Environmental Quality* 15, 13-36.

Roberts, T.M., Darrall, N.M. & Lane, P. (1984) *Advances in Applied Biology* 9, 1-142.

Roper, T.R. & Williams, L.E.,(1988) *Horticulture Science* 23, 724 (Abstr. 034).

Rosemann, D., Heller, W. & Sandermann, H. Jr. (1991) *Plant Physiology* 97, 1280-1286.

Rowland, A., Murray, A.J.S. & Wellburn, A.R. (1985) *Reviews of Environmental Health* 5, 295-342.

Sakaki, T., Kondo, N. & Sugahara, K. (1983) *Physiologia Plantarum* **59**, 28-34.

Sandermann, H., Schmitt, R., Heller, W., Rosemann, D. & Langebartels, C. (1989) In: *Acid Deposition. Sources, effects & controls*, Longhurst, J.W.S. (ed.). pp 243-254. British Library, London.

Sauter, J.J., Kammerbauer, H., Pambor, L. & Hock, B. (1987) *European Journal for Pathology* **17**, 444-448.

Saxe, H. & Murali, N.S. (1989) *Physiologia Plantarum* **76**, 349-355.

Saxe, H. (1991) *Advances in Botanical Research* **18**, 1-130.

Schaub, H., Henrich, J., Hohenberg, G. & Lenzen, B. (1990) *Staub-Reinhaltung der Luft* **50**, 241-244.

Schaub, H., Hohenberg, G., Henrich, J., Joestel, A., Knorre, U., Kuprian, M., Lenzen, B., Nitsche, I., Weil, M. & Winkler, K. (1991) *Berichte des Zentrums für Umweltforschung der Johann Wolfgang Goethe-Universität Frankfurt/M.* **14**, pp 1-227.

Schumacher, J., Loss, P. & Riesner, D. (1984) *Phytopathology* **111**, 26-36.

Shinshi, H., Mohnen, D. & Meins, F. Jr. (1987) *Proc. Natl. Acad. Sci.(USA)* **84**, 89-93.

Shinshi, H., Wenzler, H., Neuhaus, J.-M., Felix, G., Hofsteenge, J. & Meins, J.Fr. (1988) *Proc. Natl. Acad.Sci. USA* **85**, 5541-5545.

Simpson, J., Van Montagu, M. & Herrera- Estrella, L. (1986) *Science* **233**, 34-38.

Skärby, L., Troeng, E. & Boström, C.-A. (1987) *Forest Science* **33**, 801-808.

Smith, H. (1990) *Plant, Cell & Environment* **13**, 585-594.

Smith, W.H. (1990) *Air Pollution and Forests.* Interaction between air contaminants and forest ecosystems. Springer Verlag, New York.

Somssich, I.E., Schmelzer, E., Bollmann, J. & Hahlbrock, K. (1986) *Proc. Natl. Acad.Sci. USA* **83**, 2427-2430.

Srivastava, H.S., Joliffe, P.A. & Runeckles (1975) *Canadian Journal of Botany* **53**, 466-474.

Steenken, F. (1973) *Zeitschrift für Pflanzenkrankheiten und Pflanzenschutz* **80**, 513-527.

Stockhaus, J., Eckes, P., Blau, A., Schell, J. & Willmitzer, L. (1987) *Nucleic Acids Research* **15**, 3479-3491.

Sze, H. (1985) *Annual Review of Plant Physiology* **36**, 175-208.

Tanaka, K., Saji, H. & Kondo, N. (1988) *Plant Cell Physiology* **29**, 637-642.

Tanaka, K., Suda, Y., Kondo, N. & Sugahara, K. (1985) *Plant Cell Physiology* **26**, 1425-1431.

Temple, P.J. (1986) *Plant, Cell & Environment* **9**, 315-322.

Tobin, E.M. (1987) In: Phytochrome and photoregulation in plants, Furuya, M. (ed.). pp 39-40. Academic Press, Tokyo, Orlando, London.

Tsang, E.W.T., Bowler, C., Herouart, D., Van Camp, W., Villaroel, R., Genetello, C., Van Montagu, M. & Inze, D. (1991) *Plant Cell* 3, 783-792.

Van Camp, W., Bowler, C., Villaroel, R., Tsang, E.W.T., Van Montagu, M. & Inze, D. (1990) *Proc. Natl. Acad. Sci. USA* 87, 9903-9907.

Van Haut, H., Prinz, B. & Höckel, F.E. (1979) *Schriftenreihe der Landesanstalt für Immissionsschutz des Landes Nordrhein-Westfalen* , Germany 49, pp 29-65.

Wang, S.Y., Wang, C.Y. & Wellburn, A.R. (1990) In: *Stress Responses in Plants*. Adaptation and acclimation mechanisms. Alscher, R.G. (ed.). pp 147-173. Wiley-Liss Inc., New York.

Ward, E.R., Scott, S.J., Williams, S.C., Dincher, S.S., Wiederhold, D.L., Alexander, D.C., Ahl-Goy, P., Métraux, J.-P. & Ryals, J.A. (1991) *Plant Cell* 3, 1085-1094.

Wellburn, A.R. (1984) In: *Gaseous Air Pollutants and Plant Metabolism*. Proceedings of the 1st International Symposium on Air pollution and Plant Metabolism. Koziol, M.J. & Whatley, F.R. (ed.). pp 203-221. Butterworths Scientific, London.

Wellburn, A.R. (1990) *New Phytologist* 115, 395-429.

Wharton, D.C. & Weintraub, S.T. (1980) *Biochemistry & Biophysical Research Communications* 97, 236-242.

Yalpani, N., Silverman, P., Wilson, T.M.A., Kleier, D.A. & Raskin, I. (1991) *Plant Cell* 3, 809-818.

Yu, S.-W., Li, L. & Shimazaki, K.-I. (1988) *Environmental Pollution* 55, 1-13.

Ziegler, I. (1975) *Residue Review* 56, 79-105.

Zielke, H. & Sonnenbichler, J. (1990) *Naturwissenschaften* 77, 384-385.

Lotus japonicus - a model plant for structure-function analysis in nodulation and nitrogen fixation

Qunyi Jiang and Peter M. Gresshoff

Plant Molecular Genetics, Center for Legume Research, The University of Tennessee, Knoxville, TN 37901-1071

Nitrogen fixation and nodulation: importance in agriculture

While nitrogen fixation and nodulation are clearly processes of agricultural significance, one must remember that they thematically amalgamate many significant plant developmental processes. Research into nodulation and symbiotic nitrogen fixation with the *(Brady)Rhizobium* symbiont helps to clarify mechanisms of molecular signaling and plant development.

Important concepts dealing with gene regulation during differentiation, cell types development and physiology of source-sink allocations can be studied. For example, novel phytoregulatory molecules like the lipo-oligosaccharides (LPO) produced by the bacterium in response to the plant flavone signal, have been discovered (LaRouge et al, 1991), and promise to open up a totally new area of plant growth analysis (see Stacey et al, this volume). Nodulation involves interaction between plant cells themselves, and between plant and bacterial cells. An understanding of such processes at the genetic as well as biochemical level is necessary for the directed manipulation of the plant and the bacterium to produce the building blocks for an environmentally more compatible production system.

The nitrogen-fixing nodule is formed through an initiation of cell division in a precise position, suggesting the importance of plant-derived positional gradients (Rolfe and Gresshoff, 1988). Numerous cell types (transfer cells, infected and uninfected cells, nodule parenchyma and vascular tissue cells) are induced and precise source-sink relations appear to be established. For example, Hansen and Akao (1991) and others (reviewed in Caetano-Anollés and Gresshoff, 1991) demonstrated nodule initiation always opposite the xylem pole. Nodule and bacteroid biochemistry are models of carbon-nitrogen metabolism interaction. The process of nitrogen fixation itself involves some of the best understood metalloenzymes (for example, the nitrogenase subunits, uptake hydrogenase, cytochromes and cytochrome oxidase), with large information being available on their energetics and physiological regulation. The symbiosis requires the continued interaction between the plant and the microsymbiont (being the bacterium *Rhizobium* o r *Bradyrhizobium*), and plant regulatory systems themselves.

Molecular and genetic analysis of the plant genome is in progress in some plant species and has demonstrated the importance of the plant in the regulation of the developmental processes underlying cell division, cell cycle, nodule morphology and oxygen and carbon supply to the entrapped symbiont (Nap and Bisseling, 1991; Caetano-Anollés & Gresshoff, 1991; Gresshoff, 1993a,b).

While nodulation and nitrogen fixation have clear agronomic application dealing with nitrogen supplementation of the legume crop and carry-over effects to subsequent non-legume crops, the involvement of these processes are also significant to global ecology and environmental quality. If research continues to be focused strongly on model plants which are non-legumes and thus are unable to provide information about nodulation and nitrogen fixation, research into the detailed functioning and genetic manipulation of nitrogen fixation may become scare. Nodulation research effort and funding presumably will continue at present rates with soybean, alfalfa, chickpea, pea, peanut, clover, and French bean, leading to dilution of effort and funds. Duplication is unavoidable, although novel concepts may be discovered by species comparisons.. Already one finds only a limited number of laboratories working with molecular biological approaches and plant genetics in all of these legumes. Immense efforts are needed for a new researcher to establish a competitive program as significant preparatory work is needed.

Since the chosen model plant *Arabidopsis* (or maize, and tomato) is unable to provide data directly related to nodulation and nitrogen fixation, the need for a model legume is clearly substantiated.

What is the need for a model legume?

Today's research is characterized by the unifying concepts of molecular genetics, biochemistry, and cell biology. Research can become very complicated and challenging, when subtle molecular events are studied. This challenge removes the luxury to investigate every organism to the same detail. Clearly corn, tomato, rice and snapdragon (and others plant species) provided significant impetus early on for other researchers to follow. Model systems developed naturally, because they lent themselves to specific analyses and biological problems.

Many researchers are forced to work in small academic units, insufficiently large to muster the variety of disciplines needed to pursue a research problem on several levels of complexity. This led to the rapid acceptance of the concept of a model plant. This plant is the small crucifer *Arabidopsis thaliana* (the mouse ear cress; see Meyerowitz, 1987, 1990 for reviews, and Koncz et al (1992) for an outstanding and practical book on molecular genetic research methods with *Arabidopsis*). Genetic and biological features make this plant an ideal experimental organism (see Table 1), although no one ever will claim that it possesses agronomical advantages. *Arabidopsis* was easily accepted as a model plant because basic biological and genetical facts were transferable to other plant species. The utility of *Arabidopsis* lies in the ease of culture, its small diploid genome, and the willingness of researchers to share information and material. There is precedence for such approaches with *Drosophila*, *Escherichia coli*, or bacteriophage lambda.

There are some disadvantages when directing research towards one organism. Important biological phenomena may be overlooked. One such developmental process is that of bacterially induced nodulation and subsequent infection and symbiotic nitrogen fixation (see Caetano-Anollés and Gresshoff, 1991 for review). As stated above, *Arabidopsis* clearly does not have the ability to nodulate and fix nitrogen.

Why *Lotus japonicus*?

To select a plant species as a model legume, the same parameters were considered that were useful in the selection of *Arabidopsis*. The plant must have a small genome, be diploid, be small seeded, and self-fertile. It needed to grow fast, have a short generation time, and provide sufficient material for biochemical and physiological analysis. The candidate plant must be small to allow *in vitro* growth and axenic analysis of the nodulation phenotype, yet grow well enough to provide plant material for analysis. Sexual crossing must be easy (large and plenty flowers) and each successful crossing must give rise to many progeny seeds. The plant must be plastic (cultureable in many environments), must possess the ability of regeneration from tissue culture and protoplasts, and must be transformable by *Agrobacterium tumefaciens* or *A. rhizogenes*. It would be of advantage if there is a data base, and if the symbiont bacterium has microbiological and genetic advantages (i.e. plasmid-borne symbiotic genes, fast growing, antibiotic susceptibility). Certain deficiencies can be accepted if they represent shortcomings caused by absence of research rather than biological or genetic hurdles.

Table 1: Genetic and biological properties of Arabidopsis thaliana *(mouse ear cress)*

property
small genome size (about 100 megabases per haploid genome)
low chromosome number (x=n=5)
trisomics available
self-fertile, but small flowers
described ecotypes
regeneration from protoplasts and cell culture
transformation with *Agrobacterium* possible at high frequency
developmental and some auxotrophic mutants
small plant and small seed size
many seeds produced per plant (up to 10,000 possible)
short generation time (4-5 weeks from seed to seed)
axenic plant culture on agar plates or tubes
few described fungal and bacterial pathogens
infection by cauliflower mosaic virus (DNA genome)
RFLP maps
recombinant inbred RFLP and RAPD map
Yeast artificial chromosomes (YACs) carrying *Arabidopsis* sequences (about 35% ordered)
complete genome available in ordered cosmid library

Several plant candidates were considered over the last few years. Following the EMBO-Raisa meeting in Capri (October 1991), an international questionnaire was generated to focus discussion and generate consensus. Candidate species were soybean, with good data bases for both symbiotic partners; pea with excellent genetics and bacterial and physiological data bases, as well as Vicia, and chickpea. However, it was not these known species, which received the maximum number of positive responses, but two rather obscure plants. These were *Medicago truncatula* (a diploid relative of alfalfa, and capable of indeterminate (meristematic type) nodulation, and *Lotus japonicus* (forming determinate nodules similar to soybean). Interestingly, soybean finished a close third when data were expressed across all geographic regions. European respondents favored *M. truncatula* slightly over *L. japonicus*, while North American researchers, motivated by funding restrictions and the land grant university system gave soybean a slight majority over *M. truncatula* and *L. japonicus*.

The 'Australians' put up a separate rear-guard proposal suggesting that

Trifolium subterranean is an excellent candidate for research. While this plant appears to have many experimental advantages, there are no published reports on transformation. Meanwhile cynics suggested that we need a model legume tree, a model foragelegume , and a model oilseedlegume. While surveys like this are not important for future science direction, they mirror the general opinion of the science community. In the United Kingdom, research on *L. japonicus* was given a direct trust, by making it a selected model legume in government-sponsored research (AFRC). Table 2 summarizes the experimental characteristics of *L. japonicus* .

Table 2: Characteristics of Lotus japonicus *as a model legume*

Genetics	Biology
x=n=6	perennial
haploid genome size 0.5 pg	many large flowers
easy DNA isolation	large plant after 6 weeks
many inbred lines available	6 weeks to flowering
3-4 months generation time	20 seeds per pod
self-fertile	6000 seeds per plant
easy emasculation	flexible for growth
RFLP detected	regrowth from cuttings and roots
DAF[a] achieved	1.2 g per 1000 seeds
Tissue culture	**Nodulation**
protoplast regeneration	determinate nodule type
Agrobacterium tumefaciens transformation	nodulated by *Rhizobium loti* (fast growing *Rhizobium*)
plant regeneration from cell culture	root nodule appearance by 3 weeks
hygromycin and kanamycin susceptible	culture in closed tube or pouch
no endogenous GUS activity	good fixation rates

[a] *DAF = DNA amplification fingerprinting using single short arbitrarily chosen oligonucleotide primers as described by Caetano-Anollés et al (1991) BioTechnology 9:553-557. GUS= ß-glucuronidase activity; RFLP=restriction fragment length polymorphism. Information was accumulated from several sources including Dr. Jens Stougaard (Denmark).*

L. japonicus also has some experimental disadvantages. The genetics of the bacterium (*Rhizobium loti*) is primitive relative to the expensive knowledge about *R. meliloti* and *Bradyrhizobium japonicum*. One hopes that the existing bacterial probes can be used swiftly to detect similar sequence in *R. loti*. There is a need to describe the root exudates and the chemical nature of the bacterial nodulation inducing (Nod) factors. There is a need to isolate plant symbiotic mutants and candidate nodulin genes. Judging from the success by which EMS mutagenesis of legumes was used to isolate supernodulation and non-nodulation mutants (Carroll et al, 1985a,b; 1986) was applied to other legumes (e.g.. *Phaseolus vulgaris*; Park and Buttery, 1988; *Pisum sativum*, Kneen and LaRue, 1984; Jacobsen, 1984; Jacobsen and Feenstra, 1984; Messager and Duc, 1988) and other soybean lines (Akao & Kouchi, 1991; Buttery et al, 1990; Gremaud and Harper, 1989), one suspects that chemical mutagenesis is an efficient way to induce symbiotic (Nod⁻, Nts, and Fix⁻).

However, since the eventual aim of a genetic program is to detect and characterize the causative gene sequences, the alternative approach of gene tagging by insertion mutagenesis is being explored (Handsberg & Stougaard, 1992). We believe that both approaches are warranted. Chemical mutagenesis may reveal genes which, if interrupted by insertion, may have a lethal phenotype. Insertional mutations of course are easily cloned because of the molecular tag provided by the transposable element. chemically induced mutations, for which the gene product is not known, will require a more elaborate approach. This strategy is called 'positional cloning'. The clear phenotype and its mendelian segregation are used to detect a closely linked molecular marker. This may be either a restriction fragment length polymorphism (RFLP) or an amplification polymorphism generated by the recently developed single short primer genotyping methods called either DAF (Caetano-Anollés et al, 1991a) or RAPD (Williams et al, 1990). Such short primer approaches were successfully used to detect close linkage of interesting disease loci with amplification products in lettuce (Michelmore et al, 1991) or tomato (Martin et al, 1993).

The utility of positional cloning has been demonstrated in part by our work on the supernodulation (*nts*) locus of soybean. Landau-Ellis et al (1991) found a molecular RFLP clone that does not show segregation in F2 populations. The clone (pUTG-132a) has permitted the isolation of large genomic fragments with homology to the probe. Flanking molecular markers are sought using RFLP and DAF approaches. Positional cloning requires the eventual identification of a gene through sequencing of the mutant and wild-type allele. Further evidence for the identity of a cloned sequence needs to come through functional complementation of the mutant plant by the wild-type (dominant) allele. This requires plant transformation, which now is possible for soybean (see Bond and Gresshoff, this volume).

Since both insertional and chemical mutagenesis are expected to produce symbiotic mutations in *L. japonicus* in the near future, we started experiments in the direction of establishing a molecular genetic map of *Lotus japonicus*. In the following section we shall describe preliminary data on the nodulation biology of *Lotus* plants as well as the detection of amplification polymorphisms revealed by the DAF procedure.

Funding for *L. japonicus* research may be difficult, as there are no commodity-based programs. Furthermore, the attitude of referees and administrators towards a new organism tends to be negative, since funding this new system removes funds from other established research programs and biological questions. Moreover, *L. japonicus* work will be slow to take off, as many fundamental aspects, repeating technologies in other species, will be established. This generates a Catch-22 situation; many people will jump on the new system, once the basic biology is established. Even *Arabidopsis* went through several attempts to establish itself as a model plant.

Nodulation biology of *Lotus japonicus*

Several *Lotus* genotypes were made available to us by Dr. Jens Stougaard (Denmark), while *R. loti* strain NZP2037 was made available by Professor Barry Scott (Massey University, New Zealand). The symbiotic (*sym*) region in this bacterium was mapped onto the chromosome (similar to *Bradyrhizobium*) and not a symbiotic plasmid. Nodules form in either plastic growth pouches or in vermiculite. The nodules are spherical, nitrogen-fixing as determined by plant growth under nitrogen-free conditions and acetylene reduction, and took about 3 weeks to develop fully. As yet we do not understand the reasons for the slow growth of the *L. japonicus* root system and the associated slow nodulation. Is it possible that the New Zealand *Rhizobium loti* is not compatible with the Japanese plant material? Why does the plant root system grow so slow over the first 14 days? Do we lack important physiological knowledge or is this 'hard-wired'? Unless these nodulation difficulties can be overcome, *L. japonicus* will not become a practical model legume.

The *Lotus* system can be used to study rhizosphere interaction. To do so, it was decided to develop a visual assay for plant developmental stages and the presence of the bacterium. Initial morphological investigations used the hypochlorite clearing and methylene blue staining method to observe early nodulation events in *L. japonicus*. However, bacteria were not easily detected with this method, so further studies took advantage of the GUS (ß-glucuronidase) color assay and the availability of the *gus* gene on a bacterial plasmid.

As stated above, the *Lotus* plant is a fairly slow starter during the first 30 days post-germination. The primary root looked very thin compared with *G. soja*.

Accordingly, it was difficult to detect the earliest stages of nodule initiation in *L. japonicus* under lower magnification section microscopy. Fully developed, spherical nodules took up to 6 weeks to develop. Curled root hairs and some different later stages of nodule development (for example, cell division in the inner and outer cortex and an enlarged nodule meristem with isodiametric cells) were observed on 10 to 15 day post-inoculation plants grown in plastic growth pouches.

Recombinant plasmid pKW107, containing an *aph-gusA-rrnB* fusion cloned between the *Sma*I *and Bst*XI sites of Tn5 in the pSUP1021, was obtained from Dr. K. Wilson (Vienna). Conjugation of this plasmid was performed with the *mob*+ *E. coli* SM10. Recipient *R. loti* PN184, a streptomycin derivative of NZP2037 (wild-type) was obtained from Professor Barry Scott (New Zealand).

Donor and recipient cells were grown in the liquid cultures to exponential phase, mixed in Eppendorf tubes (at a ratio 1:5 in *E. coli/R. loti* matings) and concentrated by 30 seconds centrifugation. The mating mixture was carefully suspended in 0.1 ml distilled water and spread onto TY plates, and incubated at 30°C, overnight. After mating, cells were suspended and diluted in distilled water. Streptomycin and neomycin were used for counterselection and selection factors, respectively. The Nm^r Str^r transconjugants arose at a frequency of 2 X 10^{-6}. The pSUP1021 plasmid, which carried tetracycline resistance, was used as a suicide plasmid. No Tc^r Nm^r Str^r phenotypes were shown among these transconjugants. This situation led to the integration of Tn5 into the chromosome of *R. loti*. On the Nm^r Str^r plate containing GUS substrate X-Gluc (50 µg/ml), there were 55% (145/264) transconjugants showing the blue color.

Five transconjugants were inoculated onto *Lotus japonicus* B-129 "Gifu" and nodules were harvested 6 weeks post-inoculation. Bacteria isolated from these nodules maintained the Nm^r resistance and formed tiny deep-blue colonies on TY plate containing neomycin (200 µg/ml), streptomycin (200 µg/ml) and X-Gluc (50 µg/ml).

We used the expression of the *gus* gene to detect and monitor the infection process between *R. loti* and *L. japonicus*. *R. loti* transconjugants were inoculated onto the roots of *Lotus* plants 5 days post-germination. After 6 weeks, the *Lotus* roots induced by *R. loti* carrying the *gus* gene were immersed in X-Gluc assay buffer and placed in 37°C incubator overnight. Blue colonization of roots was easily observed under section microscope.

Some root and rhizosphere zones as well as curled root hairs were stained blue, indicating colonization and infection. A strong spatial distribution of *gus* expression was detected in the mature nodules 44 days post-inoculation. In these nodules blue staining was very strong, being most intense in the infected nodule interior (the site of most bacteria). Blue staining was not observed in the nodule cortex, except for some patches of blue in the nodule surface, indicating general tissue infection. By contrast, when the *Lotus* plants were inoculated by *R. loti* NZP2037 (wild-type), nodules did not develop blue color in X-Gluc assay buffer.

Breeding and genetics of *Lotus japonicus*

It was possible to hybridize *Lotus* varieties efficiently in the greenhouse using manual emasculation methods. Table 3 shows the crossing rates achieved with some lines. Line "Gifu" served as an effective female parent giving crossing frequencies around 40-50%. The reciprocal cross with "Gifu" as the male parent gave an apparently lower crossing rate.

Since the natural lines of *Lotus* have few genetic markers, and since there is a need to establish markers to verify crosses and to generate diagnostic markers for newly discovered phenotypes, we chose to look for molecular polymorphisms. Hansberg and Stougaard (1992) already detected RFLP differences between *Lotus* ecotypes. We extended this work and demonstrated amplification polymorphisms. These were generated with the single primer DNA amplification fingerprinting method (DAF; Caetano-Anollés et al, 1991, 1993), in which genomic DNA was amplified using a thermostable DNA polymerase together with single short oligonucleotide primers.

Table 3: Crossing ability of Lotus japonicus

parents	# of crosses	# of pods	# of seeds	% success
Gifu x Korea F1A	10	4	72	40
Gifu x Funakura	29	15	194	52
Gifu x Korea F1B	8	3	37	38
Korea F1B x Gifu	11	1	15	9
Gifu (control)	20	0	0	0
Korea F1B (control)	25	0	0	0

Data from 1992 season, greenhouse grown. Qunyi Jiang, forceps method. Controls were emasculated and not pollinated.

We demonstrated in separated studies that primers as short as 5 basepairs sufficed to amplify *Lotus japonicus* DNA. Primers of eight nucleotide length and a GC content between 60 and 80% were optimal, provided that primer concentration was high (10 to 30 µM was optimal) and that the template DNA concentration was low. DAF products were separated by polyacrylamide gel electrophoresis, and visualized by silver staining (Bassam et al, 1991). Several amplification polymorphisms were repeatedly detected. Most primers tested generated from 10 to 40 amplification products. Separate studies with soybean have demonstrated that these products can be used as genetic markers. About 90% of them were found to follow Mendelian segregation (R. Prabhu, unpubl. data).

Following is a brief technical outline of how the DAF patterns were generated and how the plants were grown.

Plant culture and DNA isolation:

L. japonicum lines B-129 "Gifu", B-581"Funakura", B-177 "Korea" F1A (from one plant) and F1B (from three plants) were tested. Seeds were scarified by a gentle treatment with sand paper, then sterilized in 2.7% hypochlorite 0.01% Tween 20 for 5 min, and rinsed in several changes of sterile water.

Seeds were germinated on the top of a pile of water-soaked filter paper in a Petri-dish, kept moistened by addition of some water at the side. Incubation at 25°C (in dim light) allowed seeds to swell overnight. Germinated seedlings were transferred to pots or plastic growth pouches when the rootlet was prominent. All plants were grown in the greenhouse with an 18/6 hours day/night cycle and a 24°C/18°C day/night temperature regime.

Young tissues were harvested from *Lotus* plant materials 35 days post-germination. Genomic DNA was extracted and purified according to Dellaporta et al (1983).

Protocol of DNA amplification fingerprinting

DNA amplification fingerprinting (DAF) was carried out according to Caetano-Anollés et al (1991). In short, the reaction mixture in a total volume of 25 µl contained 5 ng of template DNA, 1.5 µM 7-base or 8-base oligonucleotide primer (National Biosciences), 7.5 units of Stoffel fragment polymerase (Amplitaq Perkin-Elmer/Cetus), 200 µM of each dNTP, 10 mM Tris-HCl (pH 8.3), 50 mM KCl and 2.5 mM $MgCl_2$. The total reaction mixture was overlaid with a drop of heavy mineral oil and amplified in an Ericomp thermocycler (Ericomp Inc., San Diego, CA) for 35 cycles (using two-step cycles of 1 s at 96°C and 1 s at 30°C, heating and cooling rates of 23°C/min and 14°C/min respectively) or 33 cycles (96°C 1 s, 60°C 1 s, 72°C 2 min). Both gave identical results, so the two-step procedure was adopted. Amplification products were separated by 5% acrylamide gel, backed by a plastic Gel-Bond film (for stability and later handling) and detected by silver staining (Bassam et al, 1991).

Analysis of DNA amplification fingerprints

Ten eight nucleotide long primers (7.7a, 7.7b, 7.7c, 7.7d, 8.6d, 8.6e, 8.6f, 8.6g, 8.7f and 8.7g) and two thermocycler programs were used for *Lotus* DNA amplification. Only amplifications with primers 7.7b, 7.7c and 7.7d did not work. Primers 8.6f, 8.6g and 8.7f appeared to generate identical patterns. However, primers 7.7a, 8.6d and 8.6e generated informative fingerprint patterns which were used to discriminate the four genotypes of *L. japonicus*.

With the primer 7.7a (Figure 1, lanes 2-5), which has the nucleotide sequence 5'CGAGCTG3', DAF regenerated nine major and common amplification products at 160, 200, 290, 350, 360, 390, 450, 460 and 470 bp. "Funakura" was distinguished by two bands at 120 bp and 130 bp, which were present in all other ecotypes. "Gifu" and "KoreaF1A" showed two novel bands at 170 bp and 180bp, which were not in the other two lines ("Funakura" and "KoreaF1B"). At 190-200 bp, "Gifu", "KoreaF1A" and "KoreaF1B" showed a doublet, which was a singlet in "Funakura" (missing the lower band).

1   2   3   4   5   6   7   8   9   10   11   12   13   14

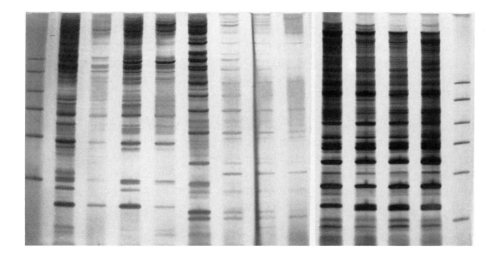

Figure 1. DNA amplification profile for Lotus japonicus *B-159 Gifu; B-581 Funakura; B-177 KoreaF1A and F1B generated with the primers 7.7a, 8.6d and 8.6e. Lane 1 and 14: DNA size markers (1000, 700, 500, 400, 300, 200 and 100 base pairs). Lane 2-5: Gifu, Funakura, KoreaF1A and KoreaF1B with primer 7.7a. Lane 6-9: Gifu, Funakura, KoreaF1A and KoreaF1B with primer 8.6e. Lane 10-13: Gifu, Funakura, KoreaF1A and KoreaF1B with primer 8.6d.*

Only "Gifu" had a strong amplification product at 210bp. In the 215-290bp range, there were 8 bands in "Gifu", compared to 4 bands in "Funakura", 5 bands in "KoreaF1A" and 6 bands in "Korea F1B". In the 300-400bp range, there are 6 bands in "Gifu", 3 bands in "Funakura" and 7 bands in "Korea F1A" or "KoreaF1B", respectively. Between 450-500bp, "Gifu" and "Funakura" showed two strong bands, while "Funakura" and "KoreaF1B" showed only weak bands.

With primer 8.6e (5'GACGTAGG3'), common bands at 140, 150, 180, 250, 330, 340, 350, 400, 500 and 700 bp were resolved. "Gifu" had unique bands at 120, 170, 270, 280 and 290bp. "Gifu" had strong band and "Funakura" had a weak band at 130bp. "Funakura", "KoreaF1A" and "KoreaF1B" had a common band at 210bp, respectively, but "Gifu" did not. At 260bp, "Funakura" and "KoreaF1A" showed a band, but this was absent in both "Gifu" and "KoreaF1B". At 270bp, "Gifu" and "Funakura" had a band, but no band at same site in "KoreaF1A" and "KoreaF1B". The term 'band' should not be confused with the term 'amplification product'. It is likely that each band contains additional overlapping or contaminating sequences. While strong amplification fragment length polymorphisms (AFLP) can be used directly as dominant genetic markers, we suggest that important markers are cloned directly from the gel and that the AFLP very likely will also represent an RFLP.

Primer 8.6d (5'GTAACGCC3') showed less variation in banding pattern. Commonality was found at 14 bands (150, 160, 180, 190, 230, 300, 350, 390, 400, 420, 510, 600, 620 and 690bp). "Funakura" had unique distinguishing bands at 210, 275 and 380 bp. "Gifu" missed bands at 225, 410 and 565bp, while "KoreaF1B" had an extra band at 430bp.

From our results, the two primers 7.7a and 8.6e were optimal for finding DNA amplification polymorphisms in *Lotus*. They showed more polymorphisms than primer 8.6d. The number of polymorphisms per primer tested is large and promises a quick analysis of the genome. We have initiated the analysis of amplification polymorphisms in the F1 and F2 progeny (see Table 3) of *L. japonicus* by the DAF technique.

Conclusions

L. japonicus was shown to have several biological and genetical advantages as a model organism. As such it will find utility in the study of environmental effects on plant growth and development. The small plant size and genome size will allow studies with large numbers. Our focus was to test the plant for its ability to work as a model legume. This aspect was rather disappointing as the plant nodulated slowly. There are some possible explanations for this and we are presently testing plant growth parameters as well as other inoculants.

The ability of the plant to be crossed was confirmed as were the molecular genetic differences between ecotypes. The DNA amplification fingerprinting (DAF) approach was applied to detect a large number of diagnostic products. These (along with others generated by these and other primers) will be followed for co-segregation in the F2 generation. Using computational linkage analysis (such as MAPMAKER) the DAF markers will be placed into linkage groups, which in turn will serve as scaffolds for morphological, developmental and biochemical markers.

Acknowledgment

This study was supported by the fund for the Ivan Racheff Chair of Excellence in Plant Molecular Genetics.

References

Akao, S. & Kouchi, H. (1992) *Soil Sci. Plant Nutr.* **38**, 183-187.

Bassam, B.J., Caetano-Anollés, G. & Gresshoff, P.M. (1991) *Anal. Biochem.* **196**, 80-83.

Buttery, B.R., Park, S.J. & Dhauvantari, B.N. (1990) *Can. J. Plant Science* **70**, 935-963.

Caetano-Anollés, G & Gresshoff, P.M. (1991) *Ann. Rev. Microbiol.* **45**, 345-382.

Caetano-Anollés, G., Bassam, B.J. & Gresshoff, P.M. (1991) *Bio/Technology* **9**, 553-557.

Caetano-Anollés, G., Bassam, B.J. & Gresshoff, P.M. (1993) (in review).

Carroll, B.J., McNeil, D.L. & P.M. Gresshoff, P.M. 1985a) *Plant Physiology* **78**, 34-40.

Carroll, B.J., McNeil, D.L. & Gresshoff, P.M. (1985b)*Proc. Natl. Acad. Sci. (USA)* **82**, 4162-4166.

Carroll, B. J., McNeil, D.L. & Gresshoff, P.M. (1986) *Plant Science* **47**, 109-119.

Chua, K.Y., Pankhurst, C.E., Macdonald, P.E., Hopcroft, D.H., Jarvis, B.D.W. & Scott, D.B. (1985) *J. Bacteriol.* **162**, 335-343.

Dellaporta, S.L., Wood, J. & Hicks, J.B. (1983) *Plant Molecular Biol. Reporter* **1**, 19-21.

Duc, G. & Messager, A. (1989) *Plant Science* **60**, 207-213.

Gremaud, M.F. & Harper, J.E. (1989)*Plant Physiol.* **89**, 169-173.

Gresshoff, P.M. & Delves, A.C. (1986) In: *Plant Gene Research III*. Blonstein, A.D. & P.J. King (eds.), p. 159-206. Springer Verlag, Wien.

Gresshoff, P.M. (1993a) In: *Nitrogen Fixation*. (eds. R. Palacios, J. Mora and W.E. Newton). Kluwer Publ. Comp. Dortrecht, The Netherlands, (in press).

Gresshoff, P. M. (1993b) *Plant Breeding Reviews* (ed. J. Janick) (in press).

Hansberg, K. & Stougaard, J. (1992) *The Plant Journal* 2, 487-496.

Hansen, A.P. & Akao, S. (1991) *J. Plant Physiol.* 138, 501-506.

Jacobsen , E. (1984) *Plant & Soil* 82, 427-438.

Jacobsen , E. & Feenstra, W.J. (1984) *Plant Sci. Lett.* B 337-344.

Kneen, B.E. & LaRue, T.A. (1984)*J. Hered.* 75, 238-240.

Koncz, C., Chua, N-H. and Schell, J. (1992) *Methods of Arabidopsis Research.* World Scientific (publ.) Singapore, New Jersey, London Hong Kong)

Landau-Ellis, D., Angermüller, S., Shoemaker, R. & Gresshoff, P.M. (1991) *Mol. Gen. Genet.* 228, 221-226.

Lerouge, P., Roche, P. Faucher, C., Maillet, F., Truchet, G., Promé, J.C., & Dénarié, J. (1990)*Nature* 244 781-784.

Martin, G. B., de Vincente, M.C. & Tanksley, S.D. (1993) *MPMI* 6, 26-34.

Meyerowitz, E.M. (1987) *Annu. Rev. Genetics* 21, 93-111.

Meyerowitz, E.M. (1990) *Trends in Genetics* 6, 1-2.

Michelmore, R.W., Paran, I. and Kesseli, R.V. (1991) *Proc. Natl. Acad. Sci. (USA)* 88, 9828-9832.

Nap, J.P. & Bisseling, T. (1990) In: *Molecular Biology of Symbiotic Nitrogen Fixation.* Gresshoff, P. M. (ed.), p. 181-229. CRC Press, Boca Raton, Fl.

Park, S.J. & Buttery, B.R. (1988) *Can. J. Plant Science* 68, 199-202.

Rolfe, B.G. & Gresshoff, P.M. (1988) *Annu. Rev. Plant Physiol. Plant Molec. Biol.* 39, 297-319.

Simon, R. (1984) *Mol. Gen. Genetics* 196, 413-420.

Williams, J.G.K., Kubelik, A.R., Livak, K.J., Rafalski & Tingey, S.V. (1990) *Nucl. Acids Research* 18, 6531-6535.

Wilson, K.J., Giller, K.E.. & Jefferson, R.A. (1991) In: *Advances in Molecular Genetics of Plant-Microbe Interaction.* Vol. 1. H.H. Hennecke and D.P.S. Verma (eds.) p. 226-229. Kluwer Academic Press (publ.). The Netherlands.

Developmental Biology of the thermophylic Cyanobacterium *Mastigocladus laminosus*

S. Edward Stevens, Jr. and Wilfredo Hernandez-Muniz

Department of Biology, Memphis State University, Memphis, TN 38152

Introduction

Mastigocladus laminosus is a cosmopolitan thermophilic *Cyanobacterium* found in thermal waters on every continent (Castenholz, 1969a). It was first described by Ferdinand Cohn as a part of the microflora of the Karlsbad hot springs, now known as Karlovy Vary, in Czechoslovakia. *M. laminosus* is a remarkably hardy organism capable of cell division and growth from 5° to about 64°C (Castenholz, 1969b; Holton, 1962; Stevens et al, 1985) and from a pH of 4.8 to 9.8 (Binder et al, 1972; Brock and Brock, 1970). It is also the most thermophilic nitrogen-fixing *Cyanobacterium* with an upper temperature limit on fixation of up to 60°C (Stewart, 1970).

In addition, *M. laminosus*, is among the most morphologically complex prokaryotic microorganisms (Balkwill et al, 1984; Hernandez-Muniz and Stevens, 1987; Nierzwicki et al, 1982; Nierzwicki-Bauer et al. 1984a,b; Stevens et al, 1985). We have observed several distinct types of differentiation in cultures of this *Cyanobacterium*. It may differentiate from narrow vegetative cells into wide vegetative cells (Nierzwicki et al, 1982). Under nitrogen-fixing conditions, both narrow and wide cells, may differentiate from vegetative cells into heterocysts (Nierzwicki-Bauer et al, 1984a,b). Typically, the heterocysts retain the general shape of the vegetative cells from which they

0-8493-8263-7/93/$0.00 + $.50

differentiate (Nierzwicki-Bauer et al, 1984b). Wide cells serve as the starting point for new narrow-celled trichomes and may also give rise to true branches in the vegetative trichome which are also of the narrow-celled morphology (Balkwill et al, 1984; Nierzwicki et al, 1982). The narrow-celled trichomes produced by the wide cells, may in turn, differentiate into motile hormogonia that apparently move by gliding motility (Castenholz, 1982; Gorbunova, 1975; Hernandez-Muniz and Stevens, 1987).

While studying the production of hormogonia we noted that a colony of actively growing trichomes of *M. laminosus* produced concentric, nearly evenly spaced rings of growth after prolonged incubation (up to 10 days) as shown in Fig. 1. Hormogonia radiated outward from the previous ring of growth by a distance of about 0.5 cm before a general cessation of movement resulted in formation of a new ring of concentrated differentiating trichomes. The rings were delineated by a zone between each one which was largely devoid of trichomes. The inner ring was darker, more intense green than the middle ring, which in turn was a more intense green than the outermost ring. The intensity of color reflected the density of trichomes in each ring. Not surprisingly, the inner ring had a larger proportion of hormogonia in an advanced state of differentiation, relative to the outermost ring. Colony formation by *M. laminosus* is not a smooth continuous process but instead occurs as a series of time-dependent pulses. It was of interest to determine what happened during and between these pulses of colony formation.

The Motile Phase of Hormogonial Development

Microscopic observations of populations of developing hormogonia of known age revealed that after 36 hr less than 5% of the total population of hormogonia were still motile. Most hormogonia had stopped moving by the end of the first 24 hr of incubation. During the motile phase hormogonia consisted exclusively of cells of the narrow cell type morphology (see also Hernandez-Muniz and Stevens, 1987). First generation hormogonia contained a profusion of intracellular granules, many of which appeared to be cyanophycin granules. During the active period of hormogonial gliding, light was not required (Hernandez-Muniz and Stevens, 1987) for motility suggesting that the energy necessary for this activity must come from reserve materials. The abundance of intracellular granules present in hormogonia would support this notion.

Cell Widening Phase

The narrow cell type observed in the hormogonium generally began to widen and round-up as gliding motility ceased. Hormogonia that had cells in the process of widening were evident after approximately 18 hr. Rarely, we observed motile hormogonia which had cells which were beginning to increase in diameter. These hormogonia moved very slowly (less than 0.5 µm/sec) probably because of the added bulk resulting from cell widening.

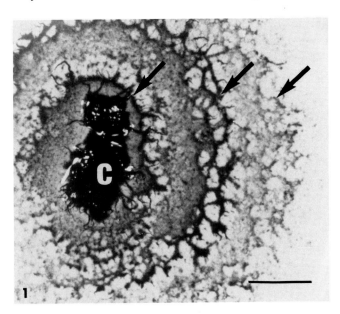

Fig. 1.   Concentric circular areas produced by successive waves of motile hormogonia. The original mass of trichomes is at the center (C). The arrows indicate the outer border of each successive wave. Bar = 0.5 cm.

The widening process of the full complement of cells making up a hormogonium began at the same time. No hormogonium has been observed which contained a mosaic of wide and narrow cells among its original cells. A possible exception was the proheterocyst, which retained the particular cell morphology (narrow or wide) it had at the onset of proheterocyst differentiation (Nierzwicki-Bauer et al, 1984b).

Light was unnecessary for motility (Hernandez-Muniz and Stevens, 1987) or for the onset of the cellular widening process, but was probably necessary for completion of the differentiation of narrow cells into wide cells. This was suggested by the results of incubating the hormogonia in the dark. After 4 days in the dark almost all of the hormogonia were senescent. Among the survivors, most had cells that were clearly in the process of widening. Completely widened cells (diameter > 6 µm) were not observed, suggesting that light, or a secondary energy source dependent on light, was necessary to complete the cell widening process.

The most likely explanation for the requirement for light at this stage is that gliding used most of the energy reserves of the hormogonium. Carbon reserves may also have become insufficient, a possibility suggested by the elevated number of carboxysomes present in wide cells relative to narrow cells (Balkwill et al, 1984). The fact that 60 - 100 new cells were produced by first generation hormogonia before mature heterocysts became apparent suggests that plentiful cellular nitrogen reserves existed to support completion of cell widening (Fig. 2). The direct requirement of light for the completion of cell widening requires further study.

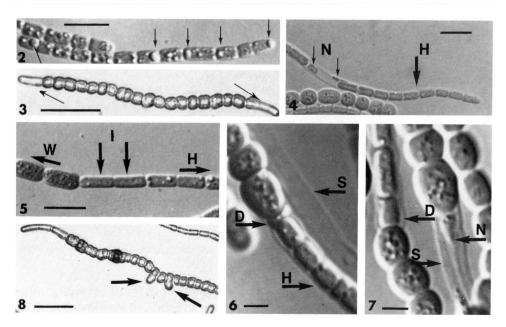

Fig. 2.   *Third generation hormogonia with cyanophycin granules. Some of the cyanophycin granules are indicated by the arrows. Bar = 10 μm*

Fig. 3.   *Developing hormogonium with wide cells and growing tips. The growing tips are indicated by the arrows. Bar = 20 μm.*

Fig. 4.   *A first generation hormogonium in the process of liberating a second generation hormogonium. The second generation hormogonium is indicated by the thick arrow and (H). Note the cell debris (N) inside the sheath indicated by the thin arrows. Bar = 10 μm.*

Fig. 5.   *Intermediate cells located between the parental wide cells and the daughter hormogonium. I = intermediate cells; W = wide cells; and, H = hormogonium. Bar = 10 μm.*

Fig. 6.   *Short, biconcave cells, similar to the "separation discs" of other species of cyanobacteria were found in the intermediate cell zone. D = biconcave cell; S = sheath; and, H = hormogonium. Bar = 3 μm.*

Fig. 7.   *Lytic debris of cell between the parental trichome and the escaping hormogonia. N=lytic debris; D = biconcave cell; and, S = sheath.   Bar = 3 μm.*

Fig. 8.   *Developing hormogonium with early branches. The early branches are indicated by the arrows. Bar = 10 μm.*

Tip Growth in Wide-Celled Trichomes

Immediately following the onset of cell widening, the cells at the tips of the trichome began to divide and reproduce (Fig. 3). This process was observed in hormogonia between 20-30 hr after release. Soon after the tips had produced about 10 cells, a new hormogonium was released (Fig. 4). In this way, a second generation of hormogonia was produced. Some hormogonia had only one growing tip.

Among the cells that comprised the growing tips, two cell types were observed: the outermost cells (narrow cell type), which formed the new hormogonium, and a new cell type, termed the intermediate cell. The outermost cells had the morphology of hormogonial cells, except that the cyanophycin granules were more abundant and larger. The intermediate cells, which connected the future hormogonium to the parental trichome, had an elongated form with very few cytoplasmic inclusions. The boundary between these two cell types was usually very clear. The intermediate cell region usually contained 2 - 8 intermediate cells (Fig. 5).

A biconcave cell usually formed close to the parental hormogonial trichome (Fig. 6 & Fig. 7) and may have been the point of separation between the second generation hormogonium and its parental trichome. Short cells with biconcave morphology, similar to the separation discs observed in other filamentous cyanobacteria (Lamont 1969), were common in the intermediate zone. It has not been determined conclusively whether or not these biconcave cells correspond to necridial cells. Cell debris and sheath material were left behind at the point of release of the new hormogonium (Figs. 6 & 7).

True Branching

Branching was never observed to start before the tip regions of the now fully wide-celled trichome had begun to grow and divide. The onset of branching was variable, but it was always observed in wide cells. The overall morphology of the branches was very similar to that of the cells in the growing tips (Fig. 8), and identical to that reported by Balkwill et al (1984) except that no oblique branching was observed.

Each branch was capable of producing at least one hormogonium during the time of observation. All wide cells may produce a branch, with the probable exception of the wide-celled heterocyst, which has never been observed to do so.

Heterocysts

The earliest time that heterocysts were distinguished in developing hormogonia was 70 hr after first release. More typical times were after 100 hr. The precise time when heterocysts became obvious varied considerably. Usually the heterocyst appeared after a period of one or two days of general yellowing of the culture. The yellowing was evident to the naked eye, but was clearer under epifluorescence microscopy because the yellowing cells gave a reduced fluorescence, suggesting that the cells were nitrogen starved (Fay, 1973; Van Gorkom and Donze, 1971). This yellowing of cells and the differentiation of heterocysts was prevented by the addition of nitrate to the medium (not shown). At the time that heterocysts became evident, a typical first generation hormogonium had produced over 60 new cells in its tips and branches. Some hormogonia produced over 100 new cells before a heterocyst became evident. Heterocysts differentiated from the vegetative cells of hormogonia of the first or later generations and appeared with both wide and narrow celled morphologies at maturity. The general morphology of the heterocyst was as described previously (Nierzwicki-Bauer et al, 1984a,b), except that the polar plugs were generally more cylindrical and larger.

It was previously established by Nierzwicki-Bauer et al (1984b), and confirmed herein, that the heterocysts of *M. laminosus* retained the shape of the vegetative cell from which it differentiated. If the differentiation of heterocysts was tied to the general pattern of development, an expected result would be the differentiation of vegetative cells into heterocysts of uniform type, either narrow or wide celled in origin. The opposite was observed.

The Developmental Cycle of *M. laminosus*

Hormogonia developed into a microcolony in a defined manner. There was some variation as to onset and duration of the different stages of development among the members of a population, but in general, the timing of discernible events in the development of hormogonia was surprisingly precise. The developmental cycle of typical hormogonia of *M. laminosus* is shown schematically in Fig. 9.

Thurston and Ingram (1971) observed the development, over time, of short trichomes of *Fischerella ambigua*, (isolated by rupturing the trichomes and plating the fragments at high dilutions) into wide-celled, branched trichomes. The differentiation of cells of the narrow cell morphology into wide cells in both *M. laminosus* and *F. ambigua* were similar.

At approximately the time that the motile hormogonia of *M. laminosus* ceased movement their further differentiation was morphologically identical to that of narrow celled vegetative trichomes. This suggests that development in *M. laminosus* follows a general pattern of differentiation with specialized patterns of differentiation for formation of the hormogonium and of the heterocyst. The accumulated evidence suggests that the general developmental pattern of *M. laminosus* follows a path, (a) from narrow celled trichomes (Nierzwicki, 1982), (b) to wide celled trichomes (Nierzwicki, 1982), (c) to a division septum parallel to the long axis of the filament forming a branch point (Balkwill et al, 1984), (d) to a tapered cell (Balkwill et al, 1984), which appears to be morphologically distinct from the intermediate cells described herein, and finally, (e) back to the narrow celled trichome. Those morphological features particular to the specialized differentiation of the hormogonium were, (a) formation of the intermediate cells, (b) accumulation of dense intracellular granules within the cells that were to become the hormogonium, (c) an apparent asymmetric cell division which produced the biconcave (necridial?) cell, (d) release of the hormogonium, and (e) the gliding of the hormogonium away from the parental trichome.

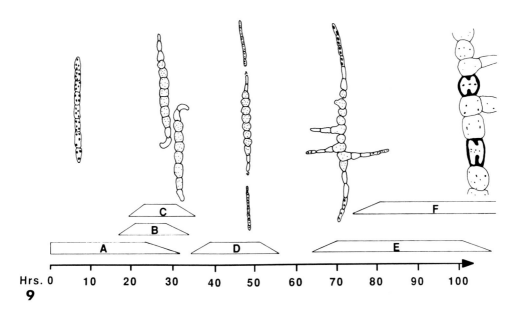

Fig. 9. Summary of events in the development of a hormogonium of M. laminosus. *A = period of gliding motility; B = period of cell widening; C = growth of hormogonial tips; D = onset of second generation of hormogonia; E = period of branching; F = period of heterocyst differentiation. Autotropism is indicated in "C". Not drawn to scale.*

Autotropism

It was observed that growing tips and branches of the hormogonia of *M. laminosus* oriented their direction of growth away from wide-celled trichomes. Autotropism was observed when a growing trichome, composed of narrow cells, was touching or growing very close to a wide-celled trichome (Figs. 10 & 11).

When two developing hormogonia were positioned in line, so that the growing tips met, the tips did not show negative tropism until they reached the area of wide cells. The result was the formation of a characteristic cross or X̲ arrangement (Fig. 11). The reorientation of growth usually involved the bending of a single cell (Fig. 10), but it occasionally involved several cells (Fig. 11). The microcolony resulting from a single hormogonium consisted of filaments of cells oriented away from the parental hormogonium (Fig. 12).

Autotropism was observed to occur at distances of up to 25 μm (Fig. 13) indicating that contact between cells was not required for it to occur.

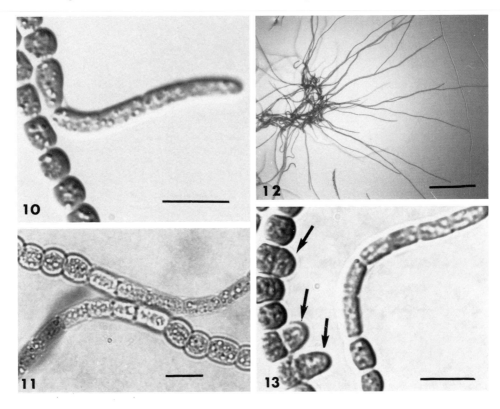

Fig. 10.   Cells of the growing tip of a trichome exhibiting autotropism. The initial cell of the tip bends sharply away from the adjacent wide celled trichome. Bar = 10 μm.

Fig. 11.   Characteristic "X" structure formed when the growing tips of two adjacent trichomes reach the wide celled part of the neighboring trichome. Note the sharp bend in each growing tip when it grew into the vicinity of the wide celled part of the opposite trichome. Bar = 10 μm.

Fig. 12.   A microcolony of M. laminosus showing the characteristic outward growth of the trichomes. The older, wide celled part of the microcolony is located in the center of the cell mass. Bar = 200 μm.

Fig. 13.   A growing tip exhibiting autotropism in the absence of direct contact with the wide cells of an adjacent trichome. Note that branch cells (arrows) occur in the same orientation. Bar = 10 μm.

Autotropism can play a key role in the establishment of the overall structure of a colony of a filamentous microorganism. Autotropism in fungal hyphae is responsible for the outward oriented growth which is characteristic of fungal colonies (Robinson, 1973). An analogous colonial structure is observed in microcolonies of *M. laminosus* (Fig. 12), and in other filamentous cyanobacteria (Lazaroff and Vishniac, 1964; Thurston and Ingram, 1971). The autotropism exhibited by *M. laminosus* is apparently unique because it is induced by only one cell type, namely wide cells. The wide cells of a developing hormogonium of *M. laminosus* are positioned in the center of the incipient microcolony.

The chemical nature of the signal inducing autotropism in *M. laminosus* is unknown. There are many reports in the literature of diffusable substances produced by different cyanobacteria (Arendarchuk, 1979; Chauhan and Gupta, 1981; Flores and Wolk, 1986, Gerasimenko and Goryunova, 1978; Hirosawa and Wolk, 1979; Kozitskaya, 1984; Lazaroff and Vishniac, 1964; Thiel and Wolk, 1982). It is tempting to speculate that substances similar to these are secreted by the wide cells of *M. laminosus* and are responsible for the autotropism observed. It is known that cultures of *M. laminosus* secrete nitrogen-containing substances (Rzhanova, 1967).

Autotropism to secreted metabolites has been implicated in the branching of filamentous fungi (Robinson, 1973). The alteration of the direction of growth observed during the autotropism of *M. laminosus* is reminiscent of the alteration of the orientation of growth that occurs during branching.

These two phenomena may be fundamentally related. The chemical basis of autotropism may also be the chemical basis of the lack of branching in narrow celled trichomes (Balkwill et al, 1984, Nierzwicki et al, 1982). It may also explain why the end cells of a wide celled hormogonium always replicate without altering the plane of division. The growing tip of such end cells is positioned at the lowest concentration of the hypothetical chemical gradient.

Acknowledgement

This study was supported by the fund for the W. Harry Feinstone Chair of Excellence in Molecular Biology.

References

Arendarchuk, V.V. (1979) *Gidrobiol. Zh.* **14**, 78-80.

Balkwill, D.L., Nierzwicki-Bauer, S.A., & Stevens, Jr., S.E. (1984) *J. Gen. Microbiology* **130**, 2079-2088.

Binder, V.A., Locher, P. & Zuber, H. (1972) *Arch. Hydrobiol.* **70**, 541-555.

Brock, T.D. & Brock, M.L. (1970) *J. Phycol.* **6**, 371-375.

Castenholz, R.W. (1969a) *Bacteriol. Rev.* **33**, 476-504.

Castenholz, R.W. (1969b) *J. Phycol.* **5**, 360-368.

Castenholz, R.W. (1982) In: *The Biology of Cyanobacteria.* Carr, N.G. & Whitton, B.A. (eds.)
    pp. 413-440. Blackwell-Scientific Publications, Oxford.

Chauhan, K.L. & Gupta, A.B. (1981) *Beitr. Biol. Pflanzen* **56**, 1-5.

Fay, P. (1973) In: *The Biology of Blue-green Algae.* Carr, N.G. & Whitton, B.A. (eds.) pp. 238-
    259. Blackwell-Scientific Publications, Oxford.

Flores, E. & Wolk, C.P. (1986) *Arch. Microbiol.* **145**, 215-219.

Gerasimenko, L.M. & Goryunova, S.V. (1978) *Mikrobiologiya* **47**, 877-880.

Gorbunova, N.P. (1975) *Vestn. Mosk. Univ. Ser. VI Biol. Pochvoved* **30**, 57-64.

Hernandez-Muniz, W. & Stevens Jr., S.E. (1987) *J. Bacteriol.* **69**, 218-223.

Hirosawa, T. & Wolk, C.P. (1979) *J. Gen. Microbiol.* **114**, 433-441.

Holton, R.W. (1962) *Amer. J. Bot.* **49**, 1-6.

Kozitskaya, V.N. (1984) *Gidrobiol. Zh.* **20**, 51-55.

Lamont, H.C. (1969) Arch. Mikrobiol. **69**, 237-259.

Lazaroff, N. & Vishniac, W. (1964) *J. Gen. Microbiol.* **35**, 447-457.

Nierzwicki, S.A., Maratea, D., Balkwill, D.L., Hardie, L.P., Mehta, V.P. & Stevens, Jr., S.E.
    (1982) *Arch. Microbiol.* **133**, 11-19.

Nierzwicki-Bauer, S.A., Balkwill, D.L. & Stevens, Jr., S.E. (1984a)*Arch. Microbiol.*
    137, 97-103.

Nierzwicki-Bauer, S.A., Balkwill, D.L., & Stevens,Jr., S.E. (1984b) *Bacteriol.* **157**, 514-525.

Robinson, P.M. (1973) *Bot. Rev.* **39**, 367-384.

Rzhanova, G.N. (1967) *Mikrobiol.* **36**, 639-645.

Stewart, W.D.P. (1970) *Phycologia* **9**, 261-268.

Stevens, Jr., S.E., Mehta, V.P. & Lane, L.S. (1985). In: *Nitrogen Fixation and CO$_2$ Metabolism.*
    Ludden, P.W., & Burris, J.E. (eds.) pp. 235-243. Elsevier Science Publishing, New York.

Thiel, T., & Wolk, C.P. (1982) *J. Phycol.* **18**, 305-306.

Thurston, E.L., & Ingram, L.O. (1971) *J. Phycol.* **7**, 203-210.

Van Gorkom, H.J., & Donze, M. (1971) *Nature* **234**, 231-232.

Mechanisms of Salt Tolerance in Cyanobacteria

Rachel Gabbay-Azaria and Elisha Tel-Or

Agricultural Botany, The Hebrew University of Jerusalem, Rehovot 76100, ISRAEL

Introduction

The ecological distribution of cyanobacteria in natural habitat of a broad range of salt concentrations, and the photoautotrophic nature of cyanobacteria make these organisms a proper model to study the mechanisms involved in salt tolerance and adaptation. Earlier studies have focused on osmoregulation in cyanobacteria, involving organic compatible solutes accumulating in the cells at steady state growth in brackish, marine and hypersaline growth media (Reed et al, 1986). Fresh water cyanobacteria, challenged with salt, demonstrated the accumulation of sucrose as compatible solute, as tested with *Nostoc muscorum* (Blumwald and Tel-Or, 1982) and *Synechococcus* 6311 (Blumwald et al, 1983). The time course of accumulation of glucosyl glycerol in *Agmenellum quadruplicatum* and its dependence on salt concentration throughout salt-upshock was demonstrating the slow response of the compatible solute biosynthesis (Tel-Or et al, 1986). We realized that in salt upshock the accumulation of compatible solute was dependent on the adaptation of the photosynthetic apparatus (Blumwald and Tel-Or, 1982) and must follow earlier responses of changes in cell volume and Na^+ translocation (Blumwald et al, 1983; Blumwald et al, 1984).

Throughout our studies on salt tolerance in *Spirulina subsalsa* we realized that it is necessary to obtain a comprehensive understanding of the events and changes during the course of cell adaptation to salt upshock, as presented schematically in Fig 1. *Spirulina subsalsa* was selected for these studies due to its salt tolerance at a concentration range between 0.25 and 2.5M NaCl (Gabbay and Tel-Or, 1985), the nature of its osmoregulation by glycinebetaine (Gabbay-Azaria et al, 1988) and the primary role of its respiration in early response to salt upshock (Gabbay-Azaria et al, 1992).

Intracellular mineral ion composition

The filamentous *Spirulina subsalsa* was grown routinely in mineral supplemented natural sea water (Gabbay and Tel-Or, 1985). The filaments formed nonhomogeneous suspension, due to aggregation with mineral precipitates. The lighter fraction of more homogenous filaments was collected for the determination of intracellular ion content during growth of batch cultures under continuous light (50 $\mu E.m^{-2}.sec^{-1}$) and stirring. As shown in Fig. 2a, Na^+ and Cl^- were the dominant ions, K^+ concentration was low, and NO_3^- and $SO_4^=$ were insignificant for osmotic consideration. Stationary phase cells, diluted in fresh sea water medium (SW), increased their Na^+ and Cl^- intracellular content within 6 to 12 hrs, and throughout this adaptation phase, the cells developed the capability to initiate Na^+ and Cl^- efflux, in the light, which was completed within 48 hrs for Cl^- and 72 hrs for Na^+. Potassium ion content in the cells was not responding to the major changes in Na^+ and Cl^- influx and efflux. Figure 2b describes the transition of cells grown in SW medium, to hypersaline medium (SW+1M NaCl) in the light. The fast entry of Na^+ and Cl^- reached maximal intracellular concentration after 12 hrs and the removal of Na^+ was initiated only 48 hrs after the salt upshock. Similarly, chlloride ion removal was carried out 48 hrs to 72 hrs after the salt upshock, demonstrating a slower mode of Na^+ and Cl^- efflux as compared to the cells in SW medium. The growth pattern of *Spirulina subsalsa* exhibited a 24 hrs lag phase in cells introduced to SW medium and a 48 hrs lag phase in cells upshocked in SW+1M NaCl medium. Hence, The efflux of Na^+ and Cl^- is suggested to be a prerequisite for the resumed biosynthetic function leading to cell growth as proposed in Figure 1.

The state of intracellular ion content in *Spirulina subsalsa* in the dark was also tested in SW and SW+1M NaCl growth media as shown in Figure 3.

The photoautotrophic cells of *Spirulina subsalsa* failed to demonstrate efflux of Na^+ and Cl^-, the influx of these ions formed a single phase in SW medium and a biphasic shape in SW+1M NaCl, and the high intracellular content of

---

opposite page: Figure 1: Problems and solutions in salt stress in Cyanobacterium *cell. Identified are physiological strategies available to adapt to salt stresses.*

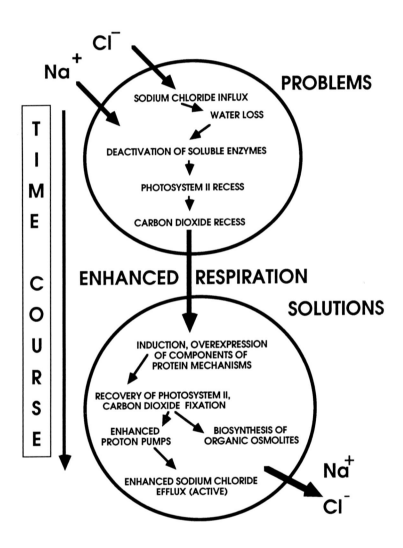

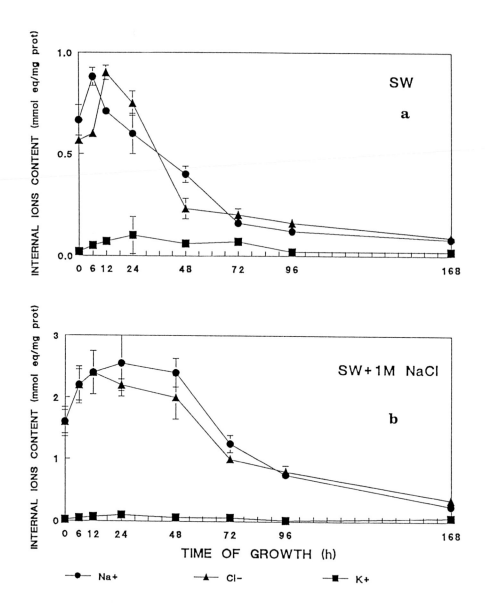

Figure 2: Intracellular ion composition in Spirulina subsalsa *grown in SW medium (a) and following salt upshock in SW+ 1M NaCl.*

Na$^+$ and Cl$^-$ was established within 96 hrs, and stayed constant throughout one week of analysis. The failure of *Spirulina subsalsa* cells to remove the intracellular salt raised uncertainty whether the cells were viable after 168 hrs in the dark. Reillumination of cultures in SW and SW+1M NaCl media, initiated photosynthesis and subsequent resumed growth after 48 hrs lag phase, suggesting that a major part of the cells maintained their viability throughout the dark experiment. Also, glycogen in the dark experiment was only partially consumed, suggesting that the cells of *Spirulina subsalsa* maintain minimal metabolic activity and do not sacrifice the carbohydrate reserves. This phenomenon of cell survival in high salt, throughout extended dark period is most striking and should be studied. Also, the possible involvement of light in the regulation of intracellular Na^+ and Cl^- content suggests possible light activation of the mechanisms leading to Na^+ and Cl^- efflux.

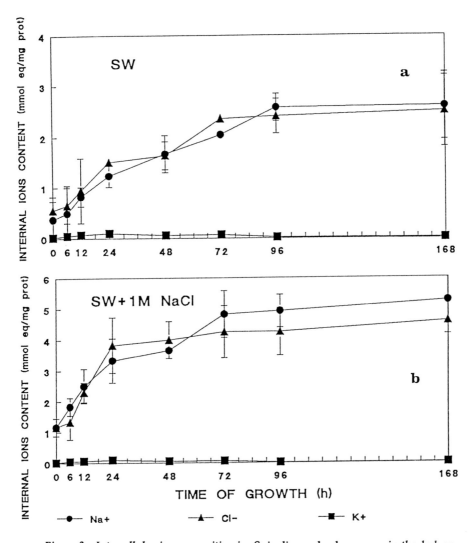

Figure 3: Intracellular ion composition in Spirulina subsalsa *grown in the dark on SW medium (a) and in SW+ 1M NaCl (b).*

The role of respiration in ion translocation

Enhanced respiration is described in Figure 1 to be driven by salt influx during salt upshock, and trigger the series of events leading to salt adaptation. Enhanced respiration, following salt upshock was reported for *Spirulina platensis* (Vonshak et al, 1988) *Synechococcus* 6311 (Fry et al, 1986) and for *Anacystis nidulans* (Paschinger, 1977; Peschek, 1987). *In vivo* respiration of *Spirulina subsalsa* cells introduced with fresh SW medium (Fig 4a) and SW +1M NaCl medium (Fig. 4b).

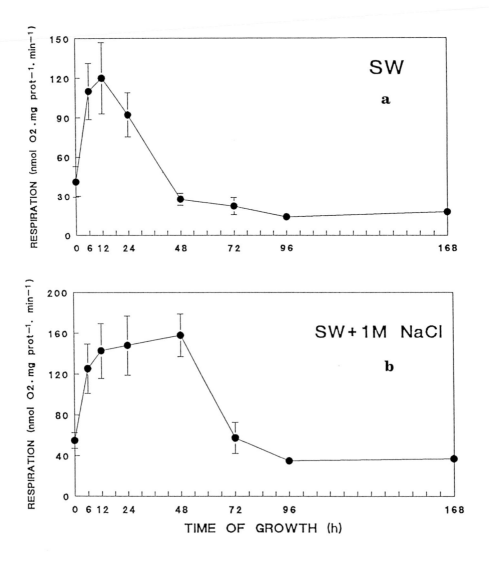

Figure 4: in vivo respiration of Spirulina subsalsa. *SW medium (a) SW+ 1M NaCl (b).*

Respiration rate of the cells follows almost identically the pattern of intracellular Na^+ and Cl^- concentration, shown above in Fig. 2a and b. The salt enhanced respiration was 3-4 fold higher in the salt upshocked cells, and was high along with the time course of enhanced Na^+ and Cl^- efflux. These results suggest that the entry of Na^+ and Cl^- trigger enhanced or activated respiratory electron transport, involving constitutive respiratory system in the cells. Recent analysis of the respiration components and their localization in *Spirulina subsalsa* (Gabbay-Azaria et al, 1992) demonstrated Na^+ enhanced respiration in protoplasts membranes, and the distribution of cytochrome oxidase in plasma membranes and thylakoid membranes. Furthermore, growth of *Spirulina subsalsa* in hypersaline medium (SW+1M NaCl) was leading to increased content of cytochrome oxidase in the plasma membranes, by salt enhanced biosynthesis.

We have tested for the possible involvement of cytochrome oxidase in proton translocation in protoplasts and plasma membranes of *Spirulina subsalsa*, and were unable to observe proton gradient formation, possibly due to the leakiness of the membranes. Alternatively, we have tested the effect of dicyclohexylcarbodiimide (DCCD), known to block proton channels in bacterial cytochrome oxidase (Pool, 1988). Figure 5 shows the effective inhibition of *Spirulina subsalsa* cytochrome oxidase activity in purified plasma membranes. These results suggest that the plasma membranes of *Spirulina subsalsa* generate throughout respiratory electron transport a proton gradient, necessary for Na^+ extrusion by the Na^+/H^+ antiport. Proton translocation cytochrome oxidase was also shown in plasma membranes of *Anacystis nidulans* (Peschek, 1984).

Activity of H^+-ATPase in enriched and purified plasma membranes of *Spirulina subsalsa* was identified and characterized. The enzyme requires Mg^{++}, has high affinity to ATP (2.5 mM), low affinity to GTP (33mM) and was inhibited by vanadate (50% erithrosin, and activity was not affected by nitrate, monensin and quabaine. A combined activity of high H^+-ATPase (50-100 μmole Pi.mg prot^{-1}.hr^{-1}) and high cytochrome oxidase (50-100 nmole cyt c oxidized.mg prot^{-1}.min^{-1}) in the purified plasma membranes of *Spirulina subsalsa* after intense capability for H^+ translocation which could drive high activity of a putative coupled Na^+/H^+ antiporter and facilitated Na^+ efflux. The pattern of enhanced respiration, shown in Figure 3, overlapping with intracellular high content of Na^+ and Cl^-, and possibly triggered or activated by high Na^+ may consist the major mechanism for Na^+ exclusion. We have not studied the nature of Cl^- exclusion, which may follow the changes in membrane potential along with Na^+ efflux, or alternatively involve specific Cl^- translocation mechanisms.

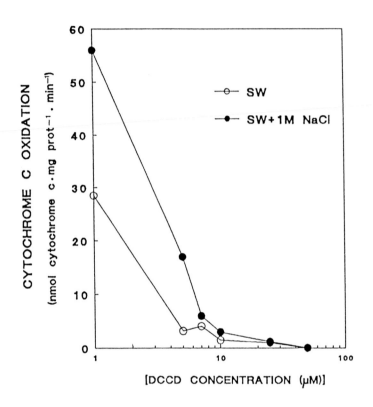

Figure 5: Cytochrome oxidase activity and its inhibition by DCCD in enriched plasma membranes of Spirulina subsulsa.

Osmoregulant accumulation and function

The biosynthesis and accumulation of organic compatible solutes in the *Cyanobacterium* cell, following salt upshock, is established as a slow and secondary response, necessary to provide a steady state osmotic stability. Biosynthesis of osmoregulant should involve the induction and expression of a number of enzyme in marine and hypersaline species where glucosyl glycerol and glycinebetaine are accumulated. These osmolites are not involved in common metabolic pathways, hence the salt stressed *Cyanobacterium* cell must sense accurately its needs for osmotic equilibrium to regulate osmotic biosynthesis in a quantitative well balanced mode. The chemical nature of compatible solutes in cyanobacteria was characterized by the introduction of ^{13}C-NMR analysis as reviewed by Reed et al (1986), and this technique was used to analyze the time course of glucosyl glycerol accumulation in *Agmenellum quadruplicatum* (Tel-Or et al, 1986) and glycinebetaine accumulation in *Spirulina subsalsa* (Gabbay-Azaria et al, 1988). Glycinebetaine biosynthesis in *Spirulina subsalsa* involves choline oxidation by betaine aldehyde dehydrogenase (Rachel Gabbay-Azaria, Ph.D. dissertation, 1992). However, we have not been able to determine the regulation of this gene by salt. We have also transformed the fresh water *Synechococcus* R-2, with the two genes for choline oxidation, but their expression was only partially affected by salt upshock, and the *Synechococcus* R-2 cells were limited by the availability of choline as precursors for glycinebetaine biosynthesis (Ofra Matan, M.Sc. dissertation, 1991).

The compatible solute is assumed to provide the *Cyanobacterium* cells with a number of protective advantages, and indeed glycinebetaine was shown to protect glutamine synthetase (Warr et al, 1984) and protect glucose-6-phosphate dehydrogenase in *Spirulina subsalsa* (Gabbay-Azaria et al, 1988).

Future objectives for salt tolerance

In conclusion we would like to list a number of open questions in the mechanisms providing salt tolerance:

a. The influence of salt upshock on photosystem 2 and the nature of its adaptation.
b. The response of ribulose bisphosphate carboxylase to salt stress and its adaptation.
c. The clarification of sodium and chloride ion extrusion.
d. Salt stress sensing by the cell, at the membrane or intracellular level.
e. Salt stress signal transduction to trigger the response to gene expression.
f. The fine tuning of compatible solute biosynthesis.

References

Blumwald, E. & Tel-Or, E. (1982) *Arch. Microbiol.* **132**, 168-172.

Blumwald, E., Melhorn, R.J. & Packer, L. (1983) *Plant Physiol.* **73**, 377-380.

Blumwald, E. Wolosin, J.M. & Packer, L. (1984) *Biochim. Biophys. Res. Comm.* **122**, 452-459.

Fry, I.V., Huflejt, M., Erber, W.W., Peschek. G.A. & Packer, L. (1986) *Arch. Biochem. Biophys.* **244**, 686-691.

Gabbay, R. & Tel-Or, E. (1985) *Plant & Soil* **89**, 107-116.

Gabbay-Azaria, R., Tel-Or, E. & Schonfeld, M. (1988) *Arch. Biochem. Biophys.* **261**, 333-339.

Gabbay-Azaria, R. & Tel-Or, E. (1991) *Biores. Tech* **38**, 215-220.

Gabbay-Azaria, R., Schonfeld, M., Tel-Or, S., Messinger, R. & Tel-Or, E. (1992)

Arch. Microbiol. **157**, 183-190.

Paschinger, H. (1977) Z. *Allgemeine Microbiol.* **17**, 377-379.

Peschek, G.A. (1984) *Plant Physiol.* **75**, 968-973.

Peschek, G.A. (1987) In: *The Cyanobacteria.* Fay, P. & Van Baalen C. (eds.) pp. 119-161. Elsevier, New York.

Pool, R.K. (1988) In: *Bacterial Energy Transduction.* Christopher, A. (ed.) pp. 231-291. Acad. Press, London.

Reed, R.H., Borowitzka, L.J., Mackay, M.A., Chudek, J.A., Foster, R., Warr, S.R.C., Moore, D.J. & Stewart, W.D.P. (1986) *FEMS Microbiol. Rev.* **39**, 51-56.

Tel-Or, E., Spath, S., Packer, L. & Melhorn, R.G. (1986) *Plant Physiol.* **82**, 646-652.

Vonshak, A., Guy, R. & Guy, M. (1988) *Arch. Microb.* **150**, 417-420.

Warr, S.R.C., Reed, R.H. & Stewart, W.P.D. (1984) *J. Gen. Microb.* **130**, 2169-2175.

Field testing transgenic crops

James E. Brandle

Research Branch, Agriculture Canada. P.O. Box 186, Delhi, Ontario, Canada N4B 2W9

Introduction

The use of transformation as a means of routine genetic modification has led to the development of new approaches to crop improvement. Some 39 crop species, ranging from vegetables to trees, have been transformed using *Agrobacterium* (Grant et al, 1991). Many other recalcitrant species have been transformed using electroporation (Rathus and Birch, 1991) or microprojectile bombardment (Franks and Birch, 1991). Genetic barriers have been lowered and a diverse array of genes and germplasm, conferring various properties, have become available for inclusion in crop varieties and will result in increased pest resistance and superior quality.

The identification of novel genes, their isolation and introduction into major crop species is a dynamic process. Field testing of the first transgenic crops is now well underway and includes a range of important dicot species. Small scale field tests are a major step towards the commercial release of transgenic crop varieties. Field testing of transgenics is controversial and proceeded slowly while various levels of government developed regulatory processes that addressed public concern and safety issues, but did not overly restrict thorough scientific evaluation. Recently, however, the number of tests has conducted in Canada (Kalous and Duke, 1989); by 1992, there were 174 field trials in progress (Agriculture Canada, 1992). A similar trend is also apparent in the United States (USDA, 1992). The increase in testing is testimony to the progress being made in the field, to improved access to the technology and to the development of the knowledge base necessary for a smooth regulatory review process. For the most part the field tests involve single genes focused on herbicide, insect and virus resistance, stress tolerance and crop quality. The purpose of this article is to document the movement of transgenic plants from the lab to the field and to review the results of the first generation of field trials.

Regulatory Issues

Regulations governing the field testing of transgenic crops in Canada were developed because the impact of this material on "human, animal and environmental health" was unknown (Kalous and Duke, 1989). The regulations were designed to minimize risk, to protect the public, but also to be sufficiently flexible to allow "relaxation in light of further experience" (Kalous and Duke, 1989).

The general feeling of groups who have studied the release of transgenic crops is that the risks of testing crops modified using molecular methods should be no different than those modified using classical breeding methods (National Research Council, 1989). A similar conclusion was reached at a meeting of the Scientific Committee on Problems of the Environment (SCOPE) and the Scientific Committee on Genetic Experimentation (COGENE), both of which are committees of the International Council of Scientific Unions. In a joint SCOPE/COGENE statement it was asserted that "most introductions of modified organisms are likely to represent low or negligible risk" (SCOPE, 1990). It was further stated that generic arguments either for or against field testing must be rejected and that each introduction should be judged individually. The Council on Scientific Affairs of the American Medical Association has indicated a need for regulation to ensure safety issues are addressed, but feels that the technology is not inherently unsafe and that the benefits far outweigh potential risk (Hendee, 1991). Guidelines recently developed for the release of transgenic organisms in the USA embody the philosophy that the products of biotechnology present no more risk to the environment than those genetically modified by traditional methods (Gavaghan, 1992).

Crawley (1990) has identified three categories of ecological risk associated with the introduction of transgenic crop plants. The first is persistence, the second is invasion and the third is transfer of introduced genes (via pollen) to other plant species. All three relate to enhanced weediness. For the most part these risks are thought to be small, albeit this conclusion has been reached in the absence of data directly related to transgenic crops. Preliminary results from the British research program, Planned Release of Selected and Modified Organisms support this position (Cherfas, 1991). Canola (*Brassica napus*) carrying genes for herbicide and kanamycin resistance was not invasive or persistent, nor did it tend to outcross with related species.

Based on food and environmental safety data provided by Calgene Inc., regulators in the U.S.A. are currently debating the safety the NPTII gene for kanamycin resistance, an important selectable marker in plants (Schaefer, 1990). In an examination of some of the issues surrounding the safety of the NPTII gene, Flavell et al (1992) concluded that there is no reason to restrict the use of this gene on grounds of safety. Aside from these initial efforts, risk assessment research is in its infancy and a number of relevant questions have yet to be answered. This work is essential to relieve public concern and to prevent the repetition of past mistakes. The data generated from the many small scale field tests being conducted around the world will, in part, answer some of the questions related to gene leakage and persistence, but detailed studies are still required. This information can then be used to educate a sceptical public, who in the end, through their elected representatives, will determine the acceptability of the products of biotechnology. The "gene law" now in effect in Germany (Kahn, 1992) is sufficient evidence of just what an uninformed and unsympathetic public can lead to and should inspire the biotechnology community to take a lead role in research related to safety.

Field Tests

Herbicide Resistance

In 1992, in Canada, 91% of the applications to field test transgenic crops involved herbicide resistance. A large number of these applications are the result of single constructs being evaluated at multiple locations. Thirty-three percent of the permits issued in the U.S. in 1992 were for field trials involving herbicide resistant transgenics. In this case a single permit may cover multiple locations. All things being equal herbicide resistance accounts for a significant proportion of the field testing that is currently being conducted. There have been two approaches to the engineering of herbicide resistance in crops: the first is the modification of the herbicide target site (usually an enzyme), either by rendering it insensitive or by inducing overproduction of an unmodified enzyme; the second method relies on the introduction of a detoxifying or degrading enzyme that attacks the herbicide (Botterman and Leemans, 1988). Resistance to seven herbicides, introduced using transformation, has been reported in 10 different crops (Duke et al, 1991).

Sulfonylurea herbicides act by inhibiting the enzyme acetohydroxyacid synthase (AHAS), which is responsible for catalyzing a common step in the synthesis of the branch chain amino acids valine, leucine and isoleucine (LaRossa and Schloss, 1984). Sulfonylurea resistant plants have been obtained by selection of haploid tobacco (*Nicotiana tabacum*) protoplasts (Chaleff and Ray, 1984) and in mutagenized *Arabidopsis thaliana* seedlings (Haughn and Sommerville, 1986). Subsequent genetic analysis of the resistant tobacco indicated that two mutant nuclear genes were involved in the expression of resistance (Chaleff and Mauvais, 1984; Chaleff and Bascomb, 1987). The herbicide resistant mutation in *Arabidopsis* was found to result from single dominant nuclear gene (Haughn and Sommerville, 1986). In both tobacco and *Arabidopsis* the mutant genes were found to co-segregate with sulfonylurea resistant AHAS activity and it was reasoned that the mutations had occurred within the AHAS gene. Two mutant genes were subsequently isolated from tobacco (*SuRB-Hra*; *SuRA-c3*) and one from *Arabidopsis* (*csr1-1*) (Lee et al, 1988; Haughn et al, 1988).

Transformation of tobacco with the *Hra* gene resulted in plants resistant to commercial rates of chlorsulfuron (Lee et al, 1988). Similar results were obtained with the *csr1-1* gene in transgenic tobacco (Haughn et al, 1988). Field tests involving five tobacco varieties transformed with the *Hra* gene were first reported in 1988 (Knowlton et al, 1988). No agronomic data were presented, but based on visual assessment the authors concluded that transgenic lines sprayed with 1, 2 and 4 times commercial rates of a sulfonylurea herbicide were unharmed. Control plants were severely damaged and showed symptoms of arrested growth, chlorosis and necrosis.

Two transgenic flax (*Linum usitatissimum*) lines harbouring the *csr1-1* gene were compared to an untransformed control for resistance to chlorsulfuron and metsulfuron methyl (McHughen and Holm, 1991). The herbicides were applied to the soil in the fall at a single rate and flax was planted the following spring. The purpose of the experiment was to evaluate the transgenic flax in a situation that simulated rotation with a sulfonylurea treated cereal. In the untreated soil, height, above ground biomass, seed yield and 1000 seed weight of the transgenic lines were found to be equal to the controls, from which the authors concluded that the introduced gene had no detrimental impact on agronomic performance. Furthermore, the transgenic lines performed similarly in herbicide treated and untreated soils indicating that transgenic flax was tolerant to the herbicide remaining in the soil at the time of planting. McHughen and Rowland (1991) conducted a multilocation evaluation, in the absence of herbicide treatments, of agronomic and seed quality characteristics of five transgenic flax lines carrying the *csr1-1* gene. There was variability in performance among the transgenic lines with the same genetic background. The authors felt that the source of the variation was either somaclonal or the result of the selection of parental plants from within a genetically variable cultivar, but was not related to the presence of T-DNA.

Brandle and Miki (1993) tested two transgenic flue-cured tobacco lines carrying the *csr1-1* gene. The sulfonylurea herbicides chlorsulfuron and DPX-R9674 were applied at two commercial rates and comparisons were made with unsprayed and untransformed controls. Both transgenic lines were lower yielding, earlier flowering, had higher sucker weights and fewer leaves than the untransformed control. The difference was attributed to untoward genetic variation introduced during the regeneration of plantlets following infection with *Agrobacterium*. One of the transgenic lines was not resistant to chlorsulfuron, the second was resistant at 10 g a.i./ha, but not at 20 g a.i./ha. Both lines were susceptible to DPX-R9674. The difference in resistance was unexpected since assays of AHAS activity had shown the enzyme to be resistant to both herbicides (Brandle et al, 1993). The authors felt that resistance to chlorsulfuron was adequate, but margins of safety needed to be increased prior to any commercial application.

The herbicides bialaphos and glufosinate are derivatives of phosphinothricin (PPT), a strong inhibitor of glutamine synthase (DeBlock et al, 1987). A PPT resistance gene (*bar*) was isolated from the bacterium *Streptomyces hygroscopicus* (Murakami et al, 1986). The resistance gene encodes an enzyme that acetylates PPT, rendering it herbicidally inactive (Thompson et al, 1987). The gene was transferred to tobacco, tomato (*Lycopersicon esculentum*) and potato (*Solanum tuberosum*), and transgenic plants showed high levels of resistance to PPT and bialaphos in greenhouse experiments (DeBlock et al, 1987).

Increases in leaf length in herbicide treated and untreated tobacco lines, transformed with the *bar* gene under the control of the 35S promoter, were compared in a field trial (De Greef et al, 1989). There was no significant decrease in leaf length at PPT rates up to 4000 g a.i./ha. PPT was also applied at two rates on three potato cultivars carrying the *bar* gene driven, in this case, by the TR2' promoter. Yield of the sprayed transgenic cultivars was not significantly different from the unsprayed, untransformed controls. Based on this information, the authors concluded that the levels of resistance in tobacco and potato were commercially acceptable. D'Halluin and co-authors (1990) evaluated 59 transgenic alfalfa (*Medicago sativa*) lines, under field conditions, for glufosinate ammonium resistance. Seventeen of the lines contained the *bar* gene under the control of the 35S promoter and the other 42 were controlled by the TR2' promoter. The authors found that the lines with the 35S promoter were significantly more tolerant than the TR2' lines and that there was variability among lines, within promoters, for expression of herbicide resistance. Unfortunately no other agronomic data were reported.

The herbicide glyphosate is non-selective and inhibits the enzyme 5-enol-pyruvylshikimate-3-phosphate synthase (EPSP), a key enzyme in the synthesis of aromatic amino acids (Rogers et al, 1983). Initial engineering efforts were two pronged. One was aimed at increasing levels of EPSP synthase by expressing a chimeric EPSP gene under the control of the 35S promoter; this effort was successful in producing glyphosate tolerant transgenic petunia (*Petunia hybrida*) plants (Shah et al, 1986). The second approach was to introduce a mutant EPSP gene, isolated from *Salmonella typhimurium*, that was insensitive to glyphosate. Tomato and tobacco plants expressing this gene showed increased tolerance to glyphosate. Initial field trials conducted with 22 transgenic tomato lines showed excellent vegetative tolerance to glyphosate, but exhibited reduced or delayed flowering relative to unsprayed controls (Fraley, 1988). Field trials involving glyphosate resistant transgenic canola (*Brassica napus*) were said to be promising (Delanney et al, 1988). New reports of field tests involving glyphosate resistant soybean (*Glycine max*) (Delanney et al, 1991) and canola (Parker et al, 1991) indicate acceptable levels of tolerance to commercial rates. Detailed results of these trials have not yet been presented.

Virus Resistance

Three approaches have been used to develop transgenic plants that are resistant to viral infection. The expression of DNA copies of viral satellite RNA and the production of large amounts of satellite RNA has been shown to greatly decrease symptom development (Harrison et al, 1987). The introduction of DNA coding for the synthesis of RNA complementary (antisense RNA) to the transcript of a coat protein gene has also been shown to increase virus protection (Cuozzo et al, 1988). Powell-Abel et al (1986) first described the use of viral coat protein (CP) genes to produce virus resistant tobacco plants. Seedlings expressing a tobacco mosaic virus (TMV) coat protein gene and inoculated with the common strain of TMV were delayed in symptom development and some failed to develop any symptoms at all. Tobacco plants expressing the TMV CP gene were also shown to be resistant to another more severe strain of TMV (Nelson et al, 1987) and to two other tobamoviruses (Nejidat and Beachy, 1990). The expression of CP genes has been the most promising of the three methods of engineering virus resistance.

Two transgenic tomato lines expressing the CP gene from the common strain of TMV showed that the transgenics were partially resistant to infection by the common strain of TMV and three strains of tomato mosaic virus (ToMV) (Nelson et al, 1988). Symptom development was also reduced. Field tests with these same lines showed that only 5% of the plants, of one of the transgenic lines, had any visual symptoms of TMV infection at the time of fruit harvest. The other line exhibited no symptoms. Virtually all of the non-transgenic control plants showed visual symptoms. Quantification of TMV accumulation corroborated visual observations, plants that expressed the CP gene and did not show symptoms, did not accumulate TMV in systemically infected leaves. Fruit yield of the inoculated transgenic lines was equal to uninoculated transgenic controls, indicating that the CP gene was providing effective protection against virus infection. Comparison of the two CP expressing lines with the untransformed check revealed that fruit yield of one of the transgenics was severely depressed. The authors attributed the low yield to some "undefined effects of plant transformation and/or regeneration". The agronomic performance of the other transgenic line was identical to the untransformed control in the absence of TMV and significantly better in the inoculated treatments. In later field trials the TMV CP gene was also shown to be effective against both the common strain and a severe TMV strain (Sanders et al, 1992). Other field trials presented in this same paper compared the TMV CP gene to the ToMV CP gene for the control of ToMV. One of the transgenic tomato lines expressing the ToMV CP gene had fewer infected plants and had lower virus titer than the transgenic line expressing the TMV CP gene. The authors concluded from this that the ToMV CP gene is a more effective means of ToMV control than the TMV CP gene. Despite substantial reductions in the incidence of infection there were no significant yield differences between inoculated and uninoculated treatments. The authors attributed this inconsistency to environmental conditions that were unconducive to symptom expression.

Transgenic potato plants expressing CP genes from two potato viruses, potato virus Y (PVY) and potato virus X (PVX), were found to be resistant to mechanical inoculation with either PVY or PVX alone and to a mix of both viruses (Lawson et al, 1990). Our transgenic lines were evaluated and one was found to maintain complete immunity to both viruses up to 36 days post-inoculation. The same line that was immune to both viruses was also immune to aphid transmission of PVY. The remaining three lines showed varying levels of susceptibility to PVY and mixed infections of PVX and PVY. Field trials designed to evaluate tuber yields and resistance to mixed PVY/PVX infections of the four transgenic lines showed that the immune line identified in the previous work was also immune under field conditions (Kaniewski et al, 1990). This line had significantly fewer infected plants than the control or any of the three remaining transgenics. Yield of the immune line was unaffected by inoculation with PVY/PVX, the remaining transgenics and the untransformed control all showed significant yield depression when inoculated. The authors concluded that CP mediated protection from mixed infections with PVY/PVX was an effective means of reducing disease development and maintaining yields in the field.

Insect Resistance

Two types of insect control agents have been introduced into plants using transformation. These are the protein delta endotoxins from *Bacillus thuringiensis* (Bt) (Vaeck et al, 1987; Fischhoff et al, 1987; Barton et al, 1987) and proteinase inhibitors (Hilder et al, 1987). The Bt system has been tested more extensively than the proteinase inhibitor system. Vaeck and coworkers (1987) isolated the *bt2* gene from *B. thuringiensis* var. *berliner* and engineered the intact bt2 gene and three truncated sequences into plant expression vectors. These were used to create transgenic tobacco plants expressing the *bt2* gene and the other truncated sequences under the control of the mannopine synthase promoter. The leaves of the transgenic plants expressing the truncated *bt2* sequences were found to be highly toxic to tobacco hornworm (*Manduca sexta*) larvae. Feeding damage to transgenic plants was substantially reduced relative to untransformed controls, indicating that the gene could effectively protect plants against insect attack.

Transgenic tomato plants expressing a truncated insect control protein gene (*Btk*) from *B. thuringiensis* var. *kurstaki* were protected from attack by tobacco hornworm, tobacco budworm (*Heliothis virescens*) and tomato fruitworm (*Heliothis zea*) larvae (Fischoff et al, 1987). Field tests involving transgenic tomato lines challenged with tobacco hornworm, tomato fruitworm and tomato pinworm (*Keiferia lycopersiclella*) were conducted for two years. Transgenic plants suffered only minor feeding damage from tobacco hornworm, while the untransformed controls were completely defoliated. With respect to hornworm damage the transgenics were also found to compare favorably to plants receiving insecticide application. Five other transgenic lines carrying the *Btk* gene, under the control of the mannopine synthase promoter, were found to vary in terms of resistance to hornworm damage. With natural infestation, damage to fruit by the tomato fruitworm was significantly lower in transgenic plants than the untransformed controls. Under heavier, artificial fruitworm infestations, damage to transgenic plants was increased, but was still significantly lower than to the controls. When difficult environmental conditions were prevalent the difference between the transgenics and the untransformed controls was much smaller. Overall, the control of tomato fruitworm was judged to be less than complete and this was attributed to the relative insensitivity of tomato fruitworm to the Bt protein. Foliar damage caused by the tomato pinworm was substantially reduced in the transgenic plants, but a high level of fruit damage was sustained. The fruit damage in the transgenic was still significantly lower than the untransformed control, but was not sufficiently effective to be commercially acceptable.

Fruit Ripening Genes

Two approaches have been taken to control fruit ripening in tomato. In the first, a bacterial gene coding for an enzyme protein that degrades the precursor to ethylene was used to transform tomato plants (Klee et al, 1991). Fruits resulting from the transgenic plants showed reduced ethylene synthesis and were significantly delayed in ripening. In the other approach, tomato plants were transformed with an antisense polygalacturonase (PG) gene. The PG enzyme is thought to play a role in pectin degradation, a process that is important to fruit softening; therefore, down regulation of the PG gene would slow fruit softening (Smith et al, 1990). Field tests involving transgenic tomatoes with a range of PG activities showed that only one line, with 8% of normal PG activity showed any significant difference from untransformed controls (Kramer et al, 1990). The percentage of rotted fruit in this line was less than in controls and juice from the fruit had greater consistency, viscosity and higher solids. Fruit yield of this line was not significantly different from the untransformed control. Of the five transgenics tested, there were two that had fruit yield that was significantly lower than the untransformed control.

Marker and Reporter Genes

Arnoldo and coworkers (1991) examined the agronomic performance of eleven transgenic canola (*Brassica napus*) lines for seed yield, oil and protein content, fatty acid profiles and glucosinolate concentration. The lines originated from two cultivars and had from one to five insertions of the neomycin phosphotransferase (NPTII) gene for kanamycin resistance. The authors compared the average performance of all the transgenic lines with that of untransformed controls and found no difference in yield or any of the seed quality characteristics. Examination of the data on a individual line basis reveals considerable variability in the performance of the transgenic lines, although no statistical analysis of this variability was presented. Hobbs et al (1990) reported variability in beta-glucuronidase (GUS) activity amongst 10 field grown transgenic tobacco lines. The transformants could be split into two groups, those with high and those with low levels of expression. Plants with low levels of expression tended to have multiple insertions of the GUS gene and increased methylation, while those with high levels had single insertions. Thornburg et al (1990) reported the expression of a chloramphenicol acetyl transferase (CAT) gene, driven by the wound-inducible promoter from the proteinase inhibitor II K gene. When insects were introduced and allowed to attack transgenic tobacco plants, the plants responded by producing CAT protein, indicating that the promoter is wound inducible under field conditions. A similar result was found when the leaves of field grown plants were wounded mechanically.

Conclusion

Most of the field testing of transgenic crops has been focused on the efficacy of the construct and has shown the genes under investigation to be stably expressed and to varying degrees, to be effective. The common thread in all the tests was variability in the expression of the transgene and, where data was provided, in agronomic performance of the transgenic lines. Agronomic variability is probably not related to the presence of the T-DNA because transgenic lines whose performance was equal to the untransformed control are present in nearly every case. This agronomic variation is most likely somaclonal in nature although there is a distinct possibility that the transferred gene has hit a plant gene of agronomic importance. Knowing that transformation is not necessarily a surgical backcross, breeders handling transgenics can act upon the variation among lines in the same way as that found in any breeding material. The lines can be selected for maximum expression of the transgene, followed by selection for agronomic and quality characteristics. Backcrossing may be necessary to fully recover the original parental phenotype. Such variation in performance serves to underscore the need for thorough evaluation of the agronomic and quality characteristics of transgenic cultivars.

Transgenic breeding lines have matured to such a stage to warrant detailed testing and are now entering the public, multilocation testing system, used to evaluate conventionally derived breeding lines from most major crops grown in Canada. Given a minimum of two years testing, assuming that side issues such as food safety are being addressed concurrently with field testing and allowing another two years for seed increase, then transgenic cultivars will be ready for release to the farming public within four years. Firm regulatory positions relative to the commercial release of transgenic crops are being developed and should established in time to fit this time frame. Transgenic cultivars are further developed in the U.S. and may be ready for release even sooner than those in Canada. Once transgenic cultivars are released, the market place will determine their success.

References

Agriculture Canada (1992) News release: 1992 Field trials approved, May 28, 1992.

Arnoldo, M., Baszczynski, C.L., Bellemare, G., Brown, G., Carlson, J., Gillespie, B., Huang, B., MacLean, N., MacRae, W.D., Rayner, G., Rozakis, S., Westecott, M. & Kemble, R.J. (1991) *Genome* 35, 58-63.

Barton, K.A., Whiteley, H.R. & Yang, N.S. (1987) *Plant Physiology* 85, 1103-1109.

Botterman, J. & Leemans, J. (1988) In: *Biotechnology and Genetic Engineering Reviews.* Russell, G.E. (ed.). pp 321-340. Intercept Ltd., Wimborne.

Brandle, J.E. & Miki, B.L. (1993) *Crop Science* (April in press).

Brandle, J.E., Labbe, H., Zilkey, B.F. & Miki, B.L. (1992) *Crop Science* 32, 1049-1053.

Chaleff, R.S. & Bascomb, N.F. (1987) *Molecular and General Genetics* 210, 33-38.

Chaleff, R.S. & Mauvais, C.J. (1984) *Science* 224, 1442-1444.

Chaleff, R.S. & Ray, T.B. (1984) *Science* 223, 1148-1151.

Cherfas, J. (1991) *Science* 251, 878.

Crawley, M.J. (1990) In: *Introduction of Genetically Modified Organisms into the Environment.* Mooney, H.A. & Bernardi, G. (eds.). pp 133-150. John Wiley & Sons, New York.

Cuozzo, M., O'Connell, K.M., Kaniewski, W., Fang, R., Chua, N. & Tumer, N.E. (1988) *Bio/Technology* **6**, 549-557.

De Block, M., Botterman, J., Vandewiele, M., Dockx, J., Thoen, C., Goessele, V., Movva, N., van Montagu, M. & Leemans, J. (1987) *European Molecular Biology Organization Journal* **6**, 2513-2518.

De Greef, W., Delon, R., De Block, M., Leemans, J. & Botterman, J. (1989) *Bio/Technology* **7**, 61-64.

Delanney, X., Metz, S.G., Lavallee, B.J., Fischhoff, D.A., Kishore, G.M., Tumer, N.E., Horsch, R.B., Rogers, S.G. & Fraley, R.T. (1988) In: *Agronomy Abstracts*. American Society of Agronomy, Madison. p 166.

Delanney, X., Lavallee, B.J., Tinius, C., Rhodes, W., Weigelt, D., Paschal E.H., Re, D.B., Barry, G.F., Eicholtz, D.A., Padgette, S.R. & Kishore, G.M. (1991) In: *Agronomy Abstracts*, American Society of Agronomy, Madison. p 194.

D'Halluin, K., Botterman, J. & De Greef, W. (1990) *Crop Science* **30**, 866-870.

Duke, S.O., Christy, A.L., Hess, F.D. & Holt, J.S. (1991) Herbicide resistant crops. Council for Agricultural Science and Technology, Ames.

Fischhoff, D.A., Bowdish, K.S., Perlak, F.J., Marrone, P.G., McCormick, S.M., Niedermeyer, J.G., Dean, D.A., Kretzmer, K.K., Mayer, E.J., Rochester, D.E., Rogers, S.G. & Fraley, R.T. (1987) *Bio/Technology* **5**, 807-813.

Flavell, R.B., Dart, E. Fuchs, R.L. & Fraley, R.T. (1992) *Bio/Technology* **10**, 141-144.

Fraley, R.T. (1988) In: *Current Communications in Molecular Biology*. Genetic improvements of agriculturally important crops, progress and issues, Fraley, R.T. & Schell, J. (eds.). pp 83-86. Cold Spring Harbour Laboratory, New York.

Franks, T. & Birch, R.G. (1991) In: *Advanced Methods in Plant Breeding and Biotechnology*, Murray, D. (ed.). pp 103-127. CAB International, Wallingford.

Gavaghan, H. (1992) *New Scientist* **133**, 10.

Grant, J.E., Dommisse, E.M., Christey, M.C. & Conner A.J. (1991) In:*Advanced Methods in Plant Breeding and Biotechnology*. Murray, D. (ed.). pp 50-73. CAB International, Wallingford.

Harrison, B.D., Mayo, M.A. & Baulcombe, D.C. (1987) *Nature* **328**, 799-802.

Haughn, G.W. & Sommerville, C. (1986) *Molecular and General Genetics* **204**, 430-434.

Haughn, G.W., Smith, J., Mazur, B. & Sommerville, C. (1988) *Molecular and General Genetics* **211**, 266-271.

Hendee, W.R. (1991) *Journal of the American Medical Association* **265**,1429-1436.

Hilder, V.A. et al. (1987) *Nature* **330**, 12-18.

Hobbs, S.L.A., Kpodar, P. & Delong, C.M.O. (1990) *Plant Molecular Biology* **15**, 851-864.
  Kahn, P. (1992) *Science* **255**, 524-526.

Kaniewski, W., Lawson, C., Sammons, B., Haley, L., Hart, J., Delanney, X. & Tumer, N.E. 1990)
  Bio/Technology **8**, 750-754.

Kalous, M.J. & Duke, L.H. (1989) *The Regulation of Biotechnology in Canada*: Part 2, The
  environmental release of genetically altered plant material. Government of Canada,
  Ottawa.

Klee, H.J., Hayford, M.B., Kretzmer, K.A., Barry, G.F. and Kishore, G.M. (1991) *The Plant Cell*
  3, 1187-1193.

Knowlton, S., Mazur, B.J. and Arntzen, C.J. (1988) In: *Current Communications in Molecular
  Biology*. Genetic improvements of agriculturally important crops, progress and issues,
  Fraley R.T. & Schell J. (eds.). pp 55-60. Cold Spring Harbour Laboratory, New York.

Kramer, M., Sanders, R.A., Sheehy, R.E., Melis, M., Kuehn, M. & Hiatt, W.R. (1990) In:
  Horticultural Biotechnology, Bennett, A.B. and O'Neill, S.D. (eds.). pp 347-355. John
  Wiley & Sons, New York.

LaRossa, R.A. and Schloss, J.V. (1984) *Journal of Biological Chemistry* **259**, 8753-8757.

Lawson, C., Kaniewski, W., Haley, L., Rozman, R., Newell, C., Sanders, P. & Tumer, N. (1990)
  Bio/Technology **8**, 127-134.

Lee, K.Y., Townsend, J., Tepperman, J., Black, M., Chui, C.F., Mazur, B., Dunsmuir, P. &
  Bedbrook, J. (1988) *European Molecular Biology Organization Journal* **7**, 1241-1248.

McHughen, A. & Holm, F. (1991) *Euphytica* **55**, 49-56.

McHughen, A., Rowland, G.G. (1991) *Euphytica* **55**, 269-275.

Murakami, T., Anzai, H., Imai, S., Satoh, A., Nagaoka, K. & Thompson, C.J. (1986) *Molecular
  and General Genetics* **205**, 42-50.

National Research Council (1989) *Field testing genetically modified organisms*: Framework for
  decisions. National Academy Press, Washington.

Nejidat, A, & Beachy, R.N. (1990) *Molecular Plant-Microbe Interactions* **3**, 247-251.

Nelson, R.S., Powell-Abel, P. & Beachy, R.N. (1987) *Virology* **158**, 126-132.

Nelson, R.S., McCormick, S.M., Delannay, X., Dube, R., Layton, J., Anderson, E.J., Kaniewska,
  M., Proksch, R.K., Horsch, R.B., Rogers, S.G., Fraley, R.T. & Beachy, R.N. (1988)
  Bio/Technology **6**, 403-409.

Parker, G.B., Mitchell, A.H., Hart, J.L., Padgette, S.R., Fedele, M.J., Barry, G.F., Didier, D.K.,
  Re, D.B., Eichholtz, D.A., Kishore, G.M. & Delanney, X. (1991) In: *Agronomy
  Abstracts*, American Society of Agronomy, Madison. p 199.

Powell-Abel, P., Nelson, R.S., Braun, D., Hoffmann, N., Rogers, S.G., Fraley, R.T. & Beachy,
  R.N. (1986) *Science* **232**, 738-743.

Rathus, C. & Birch, R.G. (1991) In: *Advanced Methods in Plant Breeding and Biotechnology*, Murray, D. (ed.). pp 74-102. CAB International, Wallingford.

Rogers, S.G., Brand, L.A., Holder, S.B., Sharps, E.A. & Bracken, M.J. (1983) *Applied and Environmental Biology* **46**, 37-43.

Sanders, P.R., Sammons, B., Kaniewski, W., Haley, W., Layton, J., LaVallee, B.J., Delanney, X. & Tumer, N.E. (1992) *Phytopathology* **82**, 683-690.

Shah, D.M., Horsch, R.B., Klee, H.J., Kishore, G.M., Winter, J.A., Tumer, N.E., Hironaka, C.M., Sanders, P.R., Gasser, C.S., Aykent, S., Siegel, N.R., Rogers, S.G. & Fraley, R.T. (1986) *Science* **233**, 478-481.

Schaefer, E. (1990) *Nature* **348**, 470.

Scientific Committee on Problems of the Environment (1990) Introduction of genetically modified organisms into the environment. John Wiley and Sons, New York.

Smith, C.J.S., Watson, C.F., Morris, P.C., Bird, C.R., Seymour, G.B., Gray, J.E., Arnold, C., Sucker, G.A., Schuch, W. Harding, S. & Grierson, D. (1990) *Plant Molecular Biology* **14**, 369-379.

Thompson, C., Movva, N., Tizard, R., Crameri, R., Davies, J., Lauwereys, M. & Botterman, J. (1987) *European Molecular Biology Organization Journal* **6**, 2519-2523.

Thornburg, R.W., Kernan, A. & Molin, L. (1990) *Plant Physiology* **92**, 500-505.

United States Department of Agriculture (1992) *Current listing of environmental release permits issued*. USDA, Hyattsville.

Vaeck, M., Reynaerts, A., Hofte, H., Jansens, S., De Beuckeleer, M., Dean, C., Zabeau, M., Van Montagu, M. & Leemans, J. (1987) *Nature* **328**, 33-37.

Role of Abscisic Acid in Plant responses to the Environment

Ma Luo[1], Robert D. Hill[1] and Shyam S. Mohapatra[1,2]

Departments of Plant Sciences[1] and Immunology[2], The University of Manitoba, Winnipeg, Manitoba R3T 2N2, Canada

Introduction

Plants are subjected to a wide variety of environmental and biological stresses throughout their life cycle, requiring their adaptation to survive conditions which are frequently harsh and variable. A large number of plant stress responses have been described which include both abiotic stress imposed by temperatures, salinity, water, drought, etc., and biotic stress imposed by wounding or infection of microorganisms.

Different stress conditions also may induce common responses such as the enhancement of phytohormones, e.g., wound induction can result in increased ethylene, auxin and abscisic acid (ABA). Despite rapid progress within the last five years in cloning and sequencing of genes having altered expression during exposure to stress, the way in which these specific changes are controlled and the molecular messages involved in that control are generally unclear.

Recent studies have demonstrated that ABA plays a cardinal role in modulation of adaptive responses of plants at the gene level under adverse environmental conditions (Chen et al, 1983; Orr et al, 1986; Ramagopal, 1987; Singh et al, 1987a; 1987b; Peña-Cortes et al, 1989). In addition to mediating the adaptive response of plants, ABA is also involved in regulation of a number of other related physiological processes which include: (i) embryo morphogenesis and the development of seeds (Eisenberg and Mascarenhas, 1985; Quatrano, 1987; Baker et al, 1988; Dure et al, 1989); (ii) seed dormancy and germination (Fong et al, 1983; Dommes and Northcote, 1985; Rodriguez et al, 1987; Koornneef et al, 1989); and (iii) plant defense from invading pathogens (Richardson et al, 1987; Dunn et al, 1990).

It has been suggested that ABA acts as a common mediator controlling adaptive plant responses to environmental stresses (Daie and Campbell, 1981) because of the numerous instances in which has been implicated. As a result of its central position, the precise role of ABA under various stress conditions is currently the subject of intense investigation in many laboratories. In this chapter, to elucidate the role of ABA plant responses to environment, we shall review (i) the studies on increase in the level of ABA after exposure to specific stress environments, (ii) the evidence of ABA-acceleration of the plant adaptation to stress, (iii) summarizing the studies on the ABA-insensitive and ABA-deficient genetic mutants, and (iv) briefly describe the regulation of expression of the ABA- and multiple environmental stress-inducible proteins and genes.

Environmental stress leads to increase in the levels of ABA

The observation that large increases occurred in the amount of ABA in leaves of plants, when they were subjected to environmental stresses indicated a possible role for ABA in plant environmental responses. Low temperatures, either above or below the freezing point, can induce stress responses in plants that will vary with species. These responses have frequently been shown to be accompanied by increases in ABA levels in the plant.

Thus, significant increases in ABA levels have been observed when tomato (*Lycopersicon esculentum* Mill. cv Venus) plants were exposed to chilling temperatures with the highest ABA levels at day/night temperatures of 10/5°C (Daie and Campbell, 1981). Water potentials did not vary in the plants during the stress treatments, suggesting that the observed ABA response was not due to temperature-induced water stress. In plants capable of withstanding freezing temperatures such as winter wheat (*Triticum aestivum* L.), potato (*Solanum commersonii)* and alfalfa (*Medicago sativa*), large increases in ABA content in leaves were observed when the plants were in the process of hardening or cold acclimation (Wightman, 1979; Chen et al, 1983; Lalk and Dorffling, 1985; Luo et al, 1992). Furthermore, it has been observed that the extent of the ABA response has a varietal dependence in winter wheat related to the hardiness of the cultivar. Thus, during acclimation, the ABA level increased more than 15-fold in the hardy winter wheat, "Kharkov", while increasing 8-fold in the moderately hardy variety "Cappelle" (Wightman, 1979). The freezing resistant variety "Holme" had higher ABA level than the less freezing resistant variety "Amandus", although the difference was not as significant (Lalk and Dorffling, 1985). Similarly, in potato, increases in ABA were observed only in *Solanum commersonii* and not in the potato species *S. tuberosum*, which failed to acclimate and was always killed at -3°C (Chen et al., 1983).

Water stress also increased the ABA content of leaves of many plants (Wright, 1978). (Zeevaart, 1980), rice (Henson, 1984), barley (Stewart and Voetberg, 1985), soybean (Bensen et al, 1988), tomato *(Lycopersicon esculentum* Mill. cv. Ailsa Craig; Bray, 1988), cotton (Hartung et al, 1988) and alfalfa *(Medicago sativa* L.; Luo et al, 1992) all have increased leaf ABA content upon exposure to water stress. Following rehydration of wilted plants or leaves, the ABA level immediately ceased to increase and returned to the pre-stress level (Zeevaart, 1980; Stewart and Voetberg, 1985; Bensen et al, 1988; Bray, 1988). This change in ABA level could be repeated in detached *Xanthium strumarium* leaves subjected to a wilting-recovery-rewilting cycle in darkness (Zeevaart, 1980).

Increases in endogenous ABA levels were also observed upon wounding. Tobacco mosaic virus (TMV) infection resulted in elevated ABA levels in young tobacco leaf tissues (Wightman, 1978). Mechanical wounding of alfalfa seedlings increased the ABA level significantly (Luo et al, 1992). Wounding of potato leaves has been shown to cause increased ABA levels and increased expression of transcripts coding for a proteinase inhibitor, while similar treatment of an ABA-deficient droopy mutant had little effect (Peña-Cortes et al, 1989). Imposition of salt stress also caused increased intracellular accumulation of ABA during tobacco cell growth (Singh et al, 1987a) and elevated endogenous ABA level in alfalfa seedlings (Luo et al, 1992).

From these selected examples, it is clear there is ample evidence in the literature to implicate ABA as a component in the response to various environmental stresses. Furthermore, comparisons of resistant and susceptible plant types suggest that the extent of production of ABA has a positive relationship with the degree of resistance acquired by the plant. These observations alone, however, are not sufficient to conclude that the hormone is a necessary intermediary in the acquisition of stress tolerance by plants.

ABA accelerates the rate of plant adaptation to stress

If ABA is a common mediator in plants for many different stresses, exogenously applied ABA should induce plant adaptive response to stress, i.e.

increase the plant stress resistance or tolerance. Several studies have shown
that exogenously applied ABA can, indeed, increase the cold tolerance of
plants. Spraying two winter wheat varieties that differ in their cold tolerance
with 10^{-4} M ABA 24 hours before a freezing test increased freezing resistance
in both varieties (Lalk and Dorffling, 1985). Exogenously applied ABA
induced frost hardiness in leaves of potato *Solanum commersonii* whether
grown under a 20°C or 2°C temperature regime (Chen et al, 1983). Applying
ABA at room temperature also increased the cold resistance of callused
Nicotiana tabacum L. explants (Bornman and Jansson, 1980), cucumber
Cucumis sativus (Flores et al, 1988), winter wheat *Triticum aestivum* L. cv
Norstar (Flores et al, 1988), and alfalfa (Mohapatra et al, 1988). It is to be
pointed out in relation to the alfalfa study that the exogenous ABA treatment
differentially affected the freezing-tolerant *M. falcata* and freezing-susceptible
M. sativa.

In some cases, however, applications of ABA were not successful in
increasing stress tolerance (Fayyaz et al, 1978). It is possible that this failure
may due to the lack of uptake, or microbial and enzyme degradation. Sterile
plant suspension cultures have been used to overcome these problems (Chen
and Gusta, 1983; LaRosa et al, 1985; Orr et al, 1986; Reaney and Gusta, 1987;
Robertson et al, 1987). Cell suspension cultures of winter wheat (*Triticum
aestivum* L. cv Norstar), winter rye (*Secale cereale* L. cv Cougar), and
bromegrass (*Bromus inermis* L.) treated with 7.5×10^{-5} molar ABA for 4 days
at 20°C could tolerate -30°C, whereas the untreated cultures tolerated only -7
to -8°C. The degree of cold-hardiness and the rate of hardening obtained by
ABA treatment was significantly higher than that induced by low
temperature alone (Chen and Gusta, 1983). When the method for the cell
suspension culture was improved, treatment of bromegrass with 7.5×10^{-5}
molar ABA for 7 days at 25°C induced hardiness beyond -60°C (Reany and
Gusta, 1987). Cell suspension cultures of winter rape (*Brassica napus* cv. *Jet
neuf*) were hardened to -20°C after treatment with 5×10^{-5} M ABA for eight
days at 25°C (Orr et al, 1986).

In addition to cold stress, ABA treatment has been reported to induce
tolerance in plants to various other stress conditions. An extreme desiccation
tolerance in the callus of resurrection plant (*Craterostigma plantagineum* H.)
was induced by treatment with ABA (5 mg/L) for at least 4 days. After the
treatment with ABA, the callus regrew following desiccation (Bartels et al.,
1990). Exogenously supplied ABA accelerated the adaptation of isolated
tobacco (*Nicotiana tabacum* L. cv. Wisconsin 38) cells to salt stress (LaRosa et
al, 1985; 1987). In another study, the role of ABA in wound-induced
expression of the proteinase inhibitor II gene in potato and tomato was
examined (Peña-Cortes et al, 1989; 1991). Potato and tomato plants
accumulated proteinase inhibitor I and II in leaves when subjected to insect
damage or mechanical wounding. These inhibitors are considered to be part

of the natural defense mechanism of plants against attacking insects (Green and Ryan, 1972; Peña-Cortes et al, 1988). Interestingly, high levels of proteinase inhibitor II gene expression were demonstrated upon exogenous application of ABA (Peña-Cortes et al, 1989).

These reports demonstrate that application of exogenous ABA elicits conditions in the plant that provide tolerance to environmental stress. Taken together with the evidence that natural adaptation to these same stresses is accompanied by an increase in endogenous ABA, there is a strong case to suggest that ABA is an intermediate in the sequence of events leading to increased tolerance to an applied environmental stress. However, again as stated previously in the case of production of ABA by exposure to stress, externally applied ABA may not be the same as those elicited by changes in endogenous levels and correlated variations may not reflect causal relationships.

ABA-insensitive and ABA-deficient mutants

Mutants are powerful tools in investigating the role of ABA in the stress tolerance of plants. ABA-deficient and/or ABA-insensitive mutants have been used to study the freezing tolerance and cold-regulated gene expression in *Arabidopsis* (Gilmour and Thomashow, 1991), the function of ABA in drought stress of tomato (Tal and Nevo, 1973; Bradford, 1983; Neil and Horgan, 1985; Bray, 1988; 1990; Cohen and Bray, 1990) and barley (Walker-Simmons et al, 1989), and the role of ABA in wound-induced expression of proteinase inhibitor II gene in potato and tomato (Peña-Cortes et al, 1989; 1991).

The development of ABA-deficient (*aba*) and ABA-insensitive (*abi*) *Arabidopsis* mutants (Koornneef et al, 1982; 1984) have been extremely useful in studying ABA function and have been used in several instances to examine the relationship between ABA and stress tolerance. The *aba* mutations affect the synthesis of ABA (Rock and Zeevaart, 1991) resulting in lowered ABA levels, while the *abi* mutants have impaired ABA sensitivity (Finkelstein and Somerville, 1990). Gilmour and Thomashow (1991) have shown that the *abi* mutations had no apparent effect on freezing tolerance, while cold-acclimated freezing tolerance was markedly impaired in the *aba* mutants compared to wild-type plants. These findings support the observations that the level of ABA during cold acclimation is important to the development of freezing tolerance but the results with the *abi* mutants leave matters unclear as to how this effect is manifested. Expression of three ABA-regulated *cor* (cold-regulated) genes was unaffected in *abi2, abi3,* and *aba1* mutants, but was greatly reduced in the *abi1* mutant, whereas cold-regulated expression of all three *cor* genes was unaffected in *abi1* mutant plants. The authors concluded that cold-regulated and ABA-regulated expression of the three *cor* genes may be mediated through independent control mechanisms (Gilmour and Thomashow, 1991), which is consistent with the findings from alfalfa (Mohapatra et al, 1988, 1989).

The ABA-deficient tomato mutant, *flacca*, is wilty as the result of an inability to close its stomata under conditions of water deficit (Tal, 1966). The mutant had 20-26% of the ABA content of the wild type when it was well-watered (Tal and Nevo, 1973; Neil and Horgan, 1985), and the wilty character could be phenotypically reversed with ABA application (Imber and Tal, 1970). Upon drought stress the leaves of *flacca* did not produce additional ABA (Neil and Horgan, 1985). By comparative 2D electrophoresis of proteins from the *flacca* mutant and the wild type several polypeptides specifically synthesized during adaptation to drought stress were detected (Bray, 1988; 1990; Cohen and Bray, 1990). Treatment of *flacca* with ABA resulted in the synthesis of these drought-stress-induced polypeptides, indicating that the synthesis of many of the polypeptides were regulated by alterations in ABA concentration during drought stress.

Another well-conducted study by Peña-Cortes et al (1989; 1991) using ABA-deficient potato and tomato mutants strongly suggested that ABA is directly involved in the induction of proteinase inhibitor II gene. Local wounding of potato or tomato plants resulted in the accumulation of proteinase inhibitors I and II throughout the aerial part of the plant (Peña-Cortes et al, 1988). ABA-deficient mutants of potato (*droopy*) and tomato (*sit*) showed a drastic reduction in expression of these genes in response to plant wounding when compared with wild-type plants. Exogenous application of ABA completely abolished the difference in the proteinase inhibitor II (PI-II) induction between wild-type and mutant plants. This demonstrated that the lack of wound response of the PI-II genes in the mutant plants was directly related to their low levels of endogenous ABA (Peña-Cortes et al, 1989). Further detailed studies using ABA-deficient mutants imply that ABA is involved in both the local and the systemic accumulation of PI-II mRNA upon wounding. ABA concentration increased both locally and in the distal unwounded leaves upon wounding and the increase was correlated with the wound induction of the PI-II gene. The potato and tomato ABA-deficient mutants did not exhibit this rise in ABA concentration and showed neither local nor systemic activation of the PI-II gene upon wounding. Exogenously applied ABA on the mutants migrated to distal, non-sprayed leaves and induced the expression of PI-II gene when the ABA concentration in this tissue was three times the control level (Peña-Cortes et al, 1991).

In essence, a general picture has emerged from the above studies, i.e., environmental stress induces an increase in ABA concentration which acts as a trigger of the defensive system of plants. Some genes are up-regulated and others are down-regulated, resulting in an overall synthesis of different genetic products which play a role in plant survival under stress.

Table 1. Stress- and ABA-inducible proteins

Clone name	Stress induced	ABA inducible	Organ specific	Function	Reference
Alfalfa					
#pUM90-1	C.S.D.W.	+	sht.	?	Luo et al (1992)
#pSM2075	C.S.D.	+	-	?	Luo et al (1991)
pSM784	C.	-	?	?	Mohapatra et
pSM2201	C.	-	?	?	al (1989)
pSM2358	C.	-	?	?	
Arabidopsis					
*cor47	C.	?	?	?	Gilmour et
@cor6.6	C.	?	?	?	al (1992)
@kin1	C.D.	+	-	?	Kurkela and Frank (1990)
Barley					
dehydrins *B8,	Di.	+	-	?	Close et al (1989)
*B9,*B17,*B18	Di.	+	-	?	
&pHVA1	C.	+	?	?	Hong et al (1988) Sutton et al (1992)
#GRP	?	?	-	?	Rohde et al (1990)
Bean					
#GRP1.8	W.	?	ov. yhc.	?	Keller et al (1988)
#GRP1.0	W.	?	ov. yhc.	?	
SAM22	W.S.	?	-	?	Crowell et al (1992)
H4	W.S.	?	-	?	
Cotton					
& Lea D7, *D11,	?	+	?	?	Baker et al (1988)
D19, D29,	?	+	?		
D34, $D113	?	+	?	?	
Chenopodium rubrum					
#GRP	L.	?	-	?	Kaldenhoff and Richter (1989)
Craterostigma plantagineum					
*pcC27-04	Di.S.	+	le. r.	?	Piatkowski et
*pcC6-19	Di.S.	+	le. r.	?	al (1990)
pcC3-06	Di.S.	+	le. r.	?	Bartels et al
pcC27-45	Di.S.	+	le. r.	?	(1990)
pcC13-62	Di.S.	+	le. r.	?	
Maize					
#pMAH9	D.W.	+	-	?	Gomez et al (1988)
dehydrin *M3	Di.	+	-	?	Close et al (1989)
*RAB-17	?	+	-	?	Vilardell et al (1990)
HS 70	H.D.W.	+	-	?	Heikkila et al (1984)

Table 1. Stress- and ABA-inducible proteins (cont.)

Clone name	Stress Induced	ABA Inducible	Organ Specific	Function	Reference
Petunia					
#petC3	W.	?	?	?	Linthorst et al (1990)
Poplar tree					
win **3**	W.	?	-	?	Bradshaw et al (1989)
Potato					
+PI-II	W.	+	-	P-ase Inh.	Peña-Cortes et al (1989)
Rape					
& Lea76	D.	+	?	?	Harada et al (1989)
Rice					
*rab 16A, *16B,	S.D.	+	-	?	Yamaguchi-Shinozaki et al (1989)
*16C, *16D	S.D.	+	-	?	Mundy and Chua (1988)
sa/T	S.D.	+	st. r.	?	Claes et al (1990)
#GRP	W.	?	?	?	Fang et al (1991)
Tobacco					
+Osmotin	S.	+	-	?	Singh et al (1989)
					Kumar and Spencer (1992)
Tomato					
+NP24	S.	?	-	?	King et al (1988)
*TAS14	S.D.	+	-	?	Godoy et al (1990)
*pLE4	D.S.C.H.	+	sht.sd	?	Cohen and Bray (1990)
pLE16	D.S.C.H.	+	sht.sd.	?	Plant et al (1991)
$pLE25	D.S.C.H.	+	sht.sd.	?	Cohen et al (1991)
+PI-II	W.	+	-	P-ase Inh.	Peña-Cortes et al (1989)
Wheat					
EM	D.	+	-	?	Marcotte et al (1989)
*rab 15	D.	+	-	?	King et al (1992)

C = cold, D = drought, Di = desiccation, H = heat, S = salt, W = wounding, ? = untested or unknown. le = leaf, ov = ovary, yhc = young hypocotyl, r = root, sd = seed, st = sheath, sht = shoot, P-ase Inh = proteinase inhibitor. Some of the data and denotations are adapted from compilations of Skriver and Mundy (1990). The symbols *, #, $, @, +, & indicate proteins share homology.

ABA- and stress-inducible proteins

Numerous studies have demonstrated that plants synthesize specific proteins upon exposure during cold acclimation (Mesa-Basso et al, 1986; Guy and

Haskell, 1987; Johnson-Flanagan and Singh, 1987; Mohapatra et al, 1987; 1988; 1989; Robertson et al, 1987; Kurkela et al, 1988a; Hahn and Walbot, 1989; Kurkela and Franck, 1990; Gilmour et al, 1992; Sutton et al, 1992). In an extensive analysis by 2-D gels of proteins isolated from a cold-tolerant and cold-susceptible alfalfa plants (Mohapatra et al, 1988), it was shown that cold acclimation induced proteins were divided into three categories: (i) those inducible by both stress and ABA, (ii) those which are specifically induced by stress but not by ABA, and (iii) those which are inducible by ABA but not by stress(es) imposed.

Similar patterns of protein synthesis have also been reported for salt stress (Ramagopal, 1987; Singh et al, 1987a; 1987b; Mundy and Chua, 1988; Singh et al, 1989; Yamaguchi-Shinozaki et al, 1989; Claes et al, 1990; Godoy et al, 1990), dehydration or desiccation stress (Baker et al, 1988; Close et al, 1989; Dure et al, 1989), water stress (Heikkila et al, 1984; Gomez et al, 1988; Bray, 1988; 1990; Cohen and Bray, 1990; Bartels et al, 1990; Piatkowski et al, 1990; Cohen et al, 1991; Plant et al, 1991; King et al, 1992), and wounding (Heikkila et al, 1984; Graham et al, 1985; Keller et al, 1988; Peña-Cortes et al, 1988; Linthorst et al, 1990; Fang et al, 1991; Crowell et al, 1992). Many of these stress-induced proteins are also induced by ABA (Heikkila et al, 1984; Singh et al, 1987a; 1989; Gomez et al, 1988; Mohapatra et al, 1988; Mundy and Chua, 1988; Close et al, 1989; Dure III et al, 1989; Peña-Cortes et al, 1989; Yamaguchi-Shinozaki et al, 1989; Bartels et al, 1990; Claes et al, 1990; Godoy et al, 1990; Kurkela and Franck, 1990; Piatkowski et al, 1990; Cohen et al, 1991; Plant et al, 1991; King et al, 1992; Luo et al, 1992). These proteins are thought to be part of the plant defense system, having some protective functions that enable plants to adapt to the adverse environment.

However, the identity and exact function of these ABA- and stress-responsive proteins are still obscure except in the case of the wound-inducible proteinase inhibitors (Peña-Cortes et al, 1989; 1991). Many of the stress- and ABA-inducible proteins have been cloned, sequenced and their expression and regulation have been studied in detail. Table 1 provides a list of these proteins according to their responses to different stresses, ABA, and structural homology. Most of the proteins listed in the table are inducible by both environmental stresses and ABA, with the exception of the cold acclimation specific genes of *M. falcata* (Mohapatra et al, 1989). To provide more information about the structure and function of these proteins, some stress-inducible proteins are also listed, even though it is not known whether they are ABA-responsive, and vise versa. Some ABA-responsive genes have been found to be stress-inducible. An example of this is the group 3 *Lea* gene *HVA1* which was induced by cold-acclimation and deacclimation (Sutton et al, 1992).

Several groups of homologous proteins listed are marked with an "*", as they are extremely hydrophilic and contained no cysteine and tryptophan. Two exceptions to this are the proteins from *Craterostigma plantagineum* and *Lea*

D-11 from cotton. Each contains one cysteine (Piatkowski et al, 1990). Proteins in this group, including most of the *rab* genes, dehydrins, some of the *Lea* genes as well as some others from a wide range of plant species, were found to be inducible by various stresses, such as desiccation, salt, drought, cold and heat. Another group of proteins (in bold face) are rich in cysteine. This group of proteins include proteinase inhibitor II, osmotin, *Lea76* and many others from different plant species. These proteins are also inducible by stresses such as wounding, salt, drought, cold, heat, and desiccation. Some were also found to be inducible by light (Kaldenhoff and Richter, 1990).

Another family of genes induced by ABA and multiple stresses encoding glycine-rich proteins have recently been described (Luo et al, 1992). These proteins were very rich in glycine (35-40%), histidine (7-15%), asparagine (8-14%) and tyrosine (5-10%). All of the encoded proteins contained characteristic tandem repeats, comprising glycine residues interrupted with histidine and/or tyrosine. Interestingly these proteins showed striking homologies with the other ABA responsive glycine-rich proteins (Gomez et al, 1988; Luo et al, 1992). It is clear from the table that a diversity of proteins was induced by ABA and/or environmental stress(es). Although these proteins are thought to have protective or adaptive functions, their exact functions and identities are as yet unknown. A subset of stress- and ABA-inducible proteins have been cloned so far, to clarify their identities and functions remains a major task.

The regulation of ABA- and stress-inducible genes

Information about nucleotide sequences of ABA- and stress-inducible genes is crucial for understanding their function and regulation. Many genes corresponding to ABA-and stress-inducible proteins have been cloned and sequenced (Table 1). To obtain a clearer understanding of the transcriptional control mechanism of ABA- and stress-induction by identifying *cis*-acting regulatory sequences, the corresponding genomic clones of some of the ABA- and stress-inducible proteins have been isolated and their nucleotide sequences have been analyzed (Baker et al, 1988; Mundy and Chua, 1988; Marcotte et al, 1989; Yamaguchi-Shinozaki et al, 1989; Plant et al, 1991). A functional analysis of the 5'-upstream region sequences of the *Em* gene of wheat identified a 50-bp ABA response element (ABRE) that was capable of conferring ABA inducibility in either orientation to a minimal cauliflower mosaic virus (CaMV) promoter (Marcotte et al, 1988; 1989). Two elements, Em1 and Em2, within this 50-bp ABRE were conserved in the promoter of other ABA-responsive genes, such as the *Lea* gene and *rab* gene families (Baker et al., 1988; Mundy and Chua, 1988; Yamaguchi-Shinozaki et al, 1989). Further study identified a DNA binding protein (EmBP-1) that interacted specifically with an 8-bp sequence in the Em1 element of the ABRE (Guiltinan et al, 1990). A 2-bp mutation (Cc**CG**g**GGC**) in the core (**ACGT**) of this sequence (C**ACGT**GGC) prevented binding of EmBP-1 and reduced the ability of the ABRE to confer ABA responsiveness on a CaMV promoter in a transient assay (Guiltinan et al, 1990). These findings suggest that the sequence C**ACGT**GGC is at least one of the ABA response elements and the DNA binding protein EmBP-1 are involved in the ABA response.

In addition, the 8-bp ABRE exactly matched the consensus G-box motif (C**ACGT**GGC), which was found in a number of yeast promoters, plant promoters regulated by visible and UV light, and in anaerobically-induced alcohol dehydrogenase promoter of *Arabidopsis* (DeLisle and Feri, 1990). The G-box is important for transcription of some of these genes. Functional analysis of the promoter region sequences of a *rab* gene also revealed the sequences conferring ABA-dependent expression on the chloramphenicol acetyl-transferase reporter gene in rice chloroplasts (Mundy et al, 1990). Two motifs in this region were found to be conserved in all *rab-16* genes. Motif I had the consensus T**ACGT**GG, which was similar to the cAMP-responsive element (TG**ACGT**CA) that bound the transcription factor CREB. The core sequence **ACGT** was conserved in this motif (Mundy et al, 1990). Motif II, which had the sequence CGSCGCGCT (S is G or C), was similar to the degenerate decanucleotide binding site of SP1, an auxiliary mammalian transcription factor. Further study showed that nuclear protein(s) bound to these motifs (Mundy et al, 1990). These data suggested that motifs I and II were possible ABREs. Comparison of the available promoter sequences of ABA- and stress-inducible genes revealed that **ACGT** cores were conserved in many promoter elements of these genes (Marcotte et al, 1989; Guiltinan et al, 1990; Mundy et al, 1990). We have summarized the promoter elements containing the **ACGT** core sequence of ABA- and stress-inducible genes in Table 2. The promoter elements containing the **ACGT** core sequence of some stress-inducible genes are also listed in the table, even though their ABA responsiveness has not been determined. The existence of the **ACGT** core sequence in the promoter region of these genes suggests that these genes may be mediated by ABA. Since previous work (Marcotte et al, 1989) has shown that a 50-bp ABA response element (ABRE) that is capable of conferring ABA inducibility in either orientation to a minimal cauliflower mosaic virus (CaMV) promoter and the sequence C**ACGT**GGC, is included in the 50-bp ABRE, we also listed the promoter elements of some stress-inducible genes containing a **TGCA** core in their sequence. It should be noted here that the promoter region of many ABA-responsive genes contain more than one sequence element with the **ACGT** core in either orientation. Whether they are all involved in the ABA or stress response remains to be tested.

Analysis of the promoter of the proteinase inhibitor II revealed that a far upstream region of the promoter (position -1300 to -700) was required for high level expression of the gene. Further deletion of the promoter to -514 abolished the low level of wound-induced expression. This proved that the activity of the proteinase inhibitor II promoter was controlled by sequences

Table 2. *Comparison of promoter elements of ABA- and stress-responsive genes in plants*

clone name	position#	sequence	response to ABA	reference
Em	-94	ACGTgcc	+	
	-149	ACGTggc		Marcotte et al (1989)
*Lea D7	-265	tACGTggt	+	
	-135	cACGTcca		
	-126	cACGTgtc		
	-108	tACGTgtt		
*Lea D11	-247	aTGCAagc	+	
*Lea D19	-153	cACGTcgc	+	
	-185	tACGTgga		
	-278	aACGTgtg		
*Lea D29	-281	cACGTacg	+	
	-189	cACGTtgc		
*Lea D34	-143	tACGTgtt	+	
	-161	cACGTgtc		
	-580	tACGTaaa		
*Lea D113	-118	tACGTggc	+	
	-139	tACGTgtc		
	-344	aACGTatt		
	-499	aACGTttc		
	-773	aACGTcaa		Baker et al (1988)
rab 16A	-146	cACGTccg	+	
	-178	tACGTgcg		
	-528	gACGTgtg		
	-982	gACGTgaa		
	-1147	gACGTtgg		Mundy and Chua (1988)
rab 16B	-60	cACGTaca	+	
	-142	cACGTacc		
	-209	cACGTccc		
	-215	tACGTaca		
	-259	tACGTggc		
	-373	cACGTgca		
	-876	tACGTgcc		
	-994	aACGTttt		
rab 16C	-24	cACGTacg	+	
	-110	cACGTacc		
	-116	cACGTaca		
	-179	cACGTcct		
	-232	tACGTggc		
	-1181	cACGTgct		Yamaguchi-Shinozaki
rab 16D	-184	tACGTggc	+	et al (1989)

upstream of position -514 (Keil et al, 1990). Further research found that two nuclear proteins specifically bound to the most 5' distal region of the promoter. DNase I protection analysis and binding to synthetic oligonucleotides identified the sequence 5'-GAGGGTATTTTCGTAA-3' as the target of these nuclear proteins (Sanchez-Serrano et al, 1990). The analysis of the promoter sequence of proteinase inhibitor IIK suggested that a regulatory element involved in the wound-inducible expression of the promoter was located in the sequence between -136 and -557. A 10-bp sequence (AAGCGTAAGT) within the region was found to bind a nuclear protein from wounded potato leaves (Palm et al, 1990).

Table 2. Comparison of promoter elements of ABA- and stress-responsive genes in plants (cont.)

clone name	position#	sequence	response to ABA	reference
rab 16D	-353	aACGTccg	+	
rab 17	-95	cACGTaca	+	
	-140	tACGTgct		
	-146	tACGTgta		
	-161	cACGTccc		Vilardell et al (1990)
pLE16	-125	cACGTaac	+	Plant et al (1991)
*win 3	-540	aACGTgta	?	
	-559	aACGTttt		Bradshaw et al (1989)
PI-IIK	-572	cACGTgga	+	Palm et al (1990)
PI-II	-767	cACGTgga	+	Keil et al (1986)
*GRP	-568	tACGTccg	?	
	-589	aACGTtca		
	-597	tACGTgtg		
	-632	tACGTgtt		
	-731	gACGTata		Rohde et al (1990)
*pMAH9	-136	cACGTccg	+	Gomez et al (1988)
*GRP1.0	-25	tTGCAttc	?	
	-125	cTGCAtgg		
	-310	aTGCAaga		
	-337	tTGCAgtg		Keller et al (1988)
*GRP1.8	-232	cTGCAtgc	?	
	-339	gTGCAaaa		
	-344	aTGCAgtg		Keller et al (1988)
*GRP	-160	aTGCAgct	?	
	-164	aTGCAtgc		
	-355	tTGCAtaa		Condit and Meagher (1986)

* *the position of the sequence is counted from the translation start site because the information about transcription start site is not available.*
# *the position of the core sequence "ACGT" or "TGCA".*
+ *ABA responsiveness has been confirmed.*
? *data are not available, or clones have not been investigated.*

Although many ABA- and stress-responsive genes have been cloned, very few of their corresponding genomic clones have been isolated. Our knowledge from limited functional analysis of the cis- and trans-acting transcription factors of a few of these genes is not enough to provide a clear picture of their regulation.

Summary and Perspectives

Taken together, the evidence so far, based on the genotypic differences with respect to (i) increased synthesis of ABA in response to stress, and (ii) acceleration of the rate of plant adaptation to stress by exogenous ABA, and the studies with the ABA-insensitive and ABA-deficient mutants have clearly revealed that ABA plays a paramount role in the process of plant adaptation to stress. At the molecular genetic level an impressive repertoire of ABA-responsive genes have been identified and the studies on the regulation of their expression have been advanced for some genes. These studies, however, have not pinpointed the precise role of ABA during stress. It is conceivable that ABA regulates the process of adaptation in two interacting steps: Firstly, ABA acts via differential signal transduction pathways on cells which are the least and the most affected by the imposed stress. Despite some efforts, a clear picture of what these pathways may be, is yet to emerge. Secondly, ABA may regulate via some genes/gene products which control the expression of stress- or adaptation-specific genes. In spite of rapid progress in analysis of several ABA responsive genes several key questions remain unanswered. For instance, does ABA interact with these genes directly? Are the ABA- and stress-specific genes coordinately regulated by the relevant cis- and trans-acting factors? What are the functions of these ABA responsive genes? Current research is focused on these and related aspects in many laboratories, the results of which are expected to unravel the molecular mechanism underlying the development of stress tolerance in plants.

Acknowledgments:

We would like to acknowledge financial support by the Natural Sciences and Engineering Research Council of Canada. We would also like to thank Sandra Kolisnyk for excellent secretarial assistance.

References

Baker, J., Steele, C. & Dure III, L. (198) *Plant Mol. Biol.* **11**, 277-291.

Bartels, D., Schneider, K., Terstappen, G., Piatkowski, D. & Salamini, F. (1990) *Planta* **181**, 27-34.

Bensen, R.J., Boyer, J.S. & Mullet, J.E. (1988) *Plant Physiol.* **88**, 289-294.

Bornman, C.H. & Jansson, E. (1980) *Physiol. Plant.* **48**, 491-493.

Bray. E.A. (1988) *Plant Physiol.* **88**, 1210-1214.

Bray, E.A. (1990) *Plant Cell Environ.* **13**, 531-538.

Bradford, K.J. (1983) *Plant Physiol.* **72**, 251-155.

Bradshaw, H.D. Jr., Hollick, J.B., Parsons, T.J., Clarke, H.R.G. & Gordon, M.P. (1989) *Plant Mol. Biol.* **14**, 51-59.

Chen, T.H.H. & Gusta, L.V. (1983) *Plant Physiol.* **73**, 71-75.

Chen, H.-H., Li, P.H. & Brenner, M.L. (1983)*Plant Physiol.* **71**, 362-365.

Claes, B., Dekeyser, R., Villarroel, R., Van den Bulcke, M., Bauw, G., Montagu, M.V. & Caplan, A. (1990) *Plant Cell* **2**, 19-27.

Close, T.J., Kortt, A.A. & Chandler, P.M. (1989) *Plant Mol. Biol.* **13**, 95-108.

Cohen, A. & Bray, E.A. (1990)*Planta* **182**, 27-33.

Cohen, A., Plant, A.L., Moses, M. & Bray, E.A. (1991) *Plant Physiol.* **97**, 1367-1374.

Condit, C.M. & Meagher, R.B. (1986) *Nature* **323**, 178-181.

Crowell, D.N., John, M.E., Russell, D. & Amasino, R.M. (1992) *Plant Mol. Biol.* **18**, 459-466.

Daie, J. & Campbell, W.F. (1981) *Plant Physiol.* **67**, 26-29.

DeLisle, A.J. & Feri, R.J. (1990) *Plant Cell* **2**, 547-557.

Dommes, J. & Northcote, D.H. (1985) *Planta* **165**, 513-521.

Dunn, R.M., Hedden, P. & Bailey, J.A. (1990) *Physiol. Mol. Plant Pathol.* **6**, 339-349.

Dure III, L., Crouch, M., Harada, J., Ho, T.-H. D., Mundy, J., Quatrano, R. Thomas, T. & Sung, Z.R. (1989) *Plant Mol. Biol.* **12**, 475-486.

Eisenberg, A.J. & Mascarenhas, J.P. (1985) *Planta* **166**, 505-514.

Fang, R.-X., Pang, Z., Gao, D.-M., Mang, K.-Q. & Chua, N.-H. (1991) *Plant Mol. Biol.* **17**, 1255-1257.

Fayyaz, M.M., McCown, B.H. & Beck, G.E. (1978) *Physiol. Plant.* **44**, 73-76.

Finkelstein, R.R. & Somerville, C.R. (1990) *Plant Physiol.* **94**, 1172-1179.

Flores, A., Grau, A., Laurich, F. & Dorffling, K. (1988)*J. Plant Physiol.* **132**, 362-369.

Fong, F., Smith, J.D. & Koehler, D.E. (1983) *Plant Physiol.* **73**, 899-901.

Gilmour, S.J. & Thomashow, M.F. (1991) *Plant Mol. Biol.* **17**, 1233-1240.

Gilmour, S.J., Artus, N.N. & Thomashow, M.F. (1992)*Plant Mol. Biol.* **18**, 13-21.

Godoy, J.A., Pardo, J.M. & Pintor-Toro, J.A. (1990) *Plant Mol. Biol.* **15**,695-705.

Gomez, J., Sanchez-Martinez, D., Stiefel, V., Rigau, J., Puigdomenech, P. & Pages, M. (1988) *Nature* **334**, 262-264

Graham, J.S., Pearce, G., Merryweather, J., Titan, K., Ericsson, L. & Ryan, C.A. (1985) *J. Biol. Chem.* **260**, 6555-6560.

Green, T.R. & Ryan, C.A. (1972) *Science* **175**, 776-777.

Guiltinan, M.J., Marcotte, W.R. & Quatrano, R. (1990) *Science* **250**, 267-270.

Guy, C.L. & Haskell, D. (1987) *Plant Physiol.* **84**, 872-878.

Hahn, M. & Walbot, V. (1989) *Plant Physiol.* **91**, 930-938.

Harada, J.J., DeLisle, A.J., Baden, C.S. & Crouch, M.l. (1989) *Plant Mol. Biol.* **12**, 395-341

Hartung, W., Radin, J.W. & Hendrix, D.L. (1988) *Plant Physiol.* **86**, 908-913.

Heikkila, J.J., Papp, J.E.T., Schultz, G.A. & Bewley, J.D. (1984) *Plant Physiol.* **76**, 270-274.

Henson, I.E. (1984) *Ann. Bot.* **54**, 569-582.

Hong, B., Uknes, S. & Ho, D.T. (1988) *Plant Mol. Biol.* **11**, 495-506.

Imber, D. & Tal, M. (1970) *Science* **169**, 592-593.

Johnson-Flanagan, A.M. & Singh, J. (1987) *Plant Physiol.* **85**, 699-705.

Kaldenhoff, R. & Richter, G. (1989) *Nucleic Acids Res.* **17**, 2853.

Keil, M., Sanchez-Serrano, J., Schell, J. & Willmitzer, L. (1986) *Nucleic Acids Res.* **14**, 5641-5650.

Keil, M., Sanchez-Serrano, J., Schell, J. & Willmitzer, L. (1990) *Plant Cell* **2**, 61-70.

Keller, B., Sauer, N. & Lamb, C.J. (1988) *EMBL J.* **7**, 3625-3633.

King, G.J., Turner, V.A., Hussey Jr., C.E., Wurtele, E.S. & Lee, S.M. (1988) *Plant Mol. Biol.* **10**, 401-412.

King, S.W., Joshi, C.P. & Nguyen, H.T. (1992) *Plant Mol. Biol.* **18**, 119-121.

Koornneef, M., Jorna, M.L., Brinkhorst-van der Swan, D.L.C. & Karssen, C.M. (1982) *Theor. Appl. Genet.* **61**, 385-393.

Koornneef, M., Reuling, G. & Karssen, C.M. (1984) *Physiol. Plant.* **61**, 377-383.

Koornneef, M., Hanhart, C.J., Hihorst, H.W.M. & Karssen, C.M. (1989) *Plant Physiol.* **90**, 462-469.

Kumar, V. & Spencer, M.E. (1992) *Plant Mol. Biol.* **18**, 621-622.

Kurkela, S. & Franck, M. (1990) *Plant Mol. Biol.* **15**, 137-144.

Kurkela, S., Franck, M., Heino, P., Lang, V. & Palva, E.T. (1988) *Plant Cell Rep.* **7,** 495-498.

Lalk, I. & Dorffling, K. (1985) *Physiol. Plant.* **63,** 287-292.

LaRosa, P.C., Handa, A.K., Hasegawa, P.M. & Bressan, R.A. (1985) *Plant. Physiol.* **79,** 138-142.

LaRosa, P.C., Hasegawa, P.M., Rhodes, D., Clithero, J.M. Watad, A.-E. A. & Bressan, R. (1987) *Plant Physiol.* **85,** 174-181.

Linthorst, H.J.M., Van Loon, L.C., Memelink, J. & Bol, J.F. (1990) *Plant Mol. Biol.* **15**, 521-523.

Luo, M., Lin, L.-H., Hill, R.D. & Mohapatra, S.S. (1991) *Plant Mol. Biol.* **17**, 1267-1269.

Luo, M., Liu, J.-H., Mohapatra, S., Hill, R.D. & Mohapatra, S.S. (1992) *J. Biol. Chem.* **267,** (in press)

Marcotte, W.R., Jr., Bayley, C.C. & Quatrano, R.S. (1988)*Nature* **335**, 454-457.

Marcotte, W.R., Jr., Russell, S.H. & Quatrano, R.S. (1989) *Plant Cell* **1,** 969-976.

Meza-Basso, L., Alberdi, M., Raynal, M., Ferrero-Cadinanos, M.-L. & Delseny, M. (1986) *Plant Physiol.* **82,** 733-738.

Mohapatra, S.S., Poole, R.J. & Dhindsa, R.S. (1987) *Plant Physiol.* **84,** 1172-1176.

Mohapatra, S.S., Poole, R.J. & Dhindsa, R.S. (1988) *Plant Physiol.* **87,** 468-473

Mohapatra, S.S., Wolfraim, L., Poole, R.J. & Dhindsa, R.S. (1989) *Plant Physiol.* **89,** 375-380

Mundy, J. & Chua, N.-H. 1988 *EMBO J.* **7,** 2279-2286

Mundy, J., Yamaguchi-Shinozaki, K. & Chua, N.-H. (1990) *Proc. Natl. Acad. Sci. USA* **87,** 1406-1410

Neill, S.J. & Horgan, R. (1985) *J. Exp. Bot.* **36,** 1222-1231.

Orr, W., Keller, W.A. & Singh, J. (1986) *J. Plant Physiol.* **126,** 23-32.

Palm, C.J., Costa, M.A., An, G. & Ryan, C.A. (1990) *Proc. Natl. Acad. Sci. USA* **87,** 603-607.

Peña-Cortes, H., Sanchez-Serrano, J.J., Rocha-Sosa, M. & Willmitzer, L. (1988) *Planta* **174,** 84-89.

Peña-Cortes, H., Sanchez-Serrano, J., Mertens, R., Willmitzer, L. & Prat, S. (1989) *Proc. Natl. Acad. Sci. USA* **86,** 9851-9855.

Peña-Cortes, H., Willmitzer, L. & Sanchez-Serrano, J.J. (1991) *Plant Cell* **3,** 963-972

Piatkowski, D., Schneider, K., Salamini, F. & Bartels, D. (1990) *Plant Physiol.* **94**, 1682-1688.

Plant. A.L., Cohen, A., Moses, M.S. & Bray, E.A. (1991) *Plant Physiol.* **97**, 900-906.

Quatrano, R.S. (1987) In: *Plant Hormones and Their Role in Plant Growth and Development*, Davies, P.J. (ed.). pp. 495-514. Dordrecht/Kluwer Academic Publishing.

Ramagopal, S. (1987) *Proc. Natl. Acad. Sci. U.S.A.* **84**, 94-98.

Reaney, M.J.T. & Gusta, L.V. (1987) *Plant Physiol.* **83**, 423-427.

Richardson, M., Valdez-Rodriguez, S. & Blanco-Labra, A. (1987) *Nature* **327**, 432-434.

Robertson, A.J., Gusta, L.V., Reaney, M.J.T. & Ishikawa, M. (1987) *Plant Physiol.* **84**, 1331-1336.

Rock, C.D. & Zeevaart, J.A.D. (1991) *Proc. Natl. Acad. Sci. USA* **88**, 7496-7499.

Rodriguez, D., Dommes, J. & Northcote, D.H. (1987) *Plant Mol. Biol.* **9**, 227-235.

Rohde, W., Rosch, K., Kroger, K. & Salamini, F. (1990) *Plant Mol. Biol.* **14**, 1057-1059.

Sanchez-Serrano, J.J., Peña-Cortes, H., Willmitzer, L. & Prat, S. (1990) *Proc. Natl. Acad. Sci. USA* **87**, 7205-7209.

Showalter, A.M., Bell, J.N., Cramer, C.L., Bailey, J.A., Varner, J.E. & Lamb, C.J. (1985) *Proc. Natl. Acad. Sci. USA* **82**, 6551-6555.

Singh, N.K., LaRosa, P. C., Hansa, A.K., Hasegawa, P.M. & Bressan, R.A. (1987a) *Proc. Natl. Acad. Sci. USA* **84**, 739-743.

Singh, N.K., Bracker, C.A., Hasegawa, P.M., Handa, A.K., Buckel, S., Hermodson, M.A., Pfankoch, E., Regnier, F.E. & Bressan, R.A. (1987b) *Plant Physiol.* **85**, 529-536.

Singh, N.K., Nelson, D.E., Kuhn, D., Hasegawa, P.M. & Bressan, R.A. (1989) *Plant Physiol.* **90**, 1096-1101.

Skriver, K. & Mundy, J. (1990) *Plant Cell* **2**, 503-512.

Stewart, C.R. & Voetberg, G. (1985) *Plant Physiol.* **79**, 24-27.

Sutton, F., Ding, X. & Kenefick, D.G. (1992) *Plant Physiol.* **99**, 338-340.

Tal, M. (1966) *Plant Physiol.* **41**, 1387-1391.

Tal, M. & Nevo, Y. (1973) *Biochem. Genet.* **8**, 291-300.

Vilardell, J., Goday, A., Freire, M.A., Torrent, M. Martinez, M.C., Torne, J.M. & Pages, M. (1990) *Plant Mol. Biol.* **14**, 423-432.

Walker-Simmons, M., Kudrna, D.A. & Warner, R.L. (1989) *Plant Physiol.* **90**, 728-733.

Wightman, F. (1979) In: *Plant Regulation and World Agriculture*. Scott, T.K. (ed.). pp 327-377. Plenum Press, New York.

Wright, S.T.C. (1978) In: *Phytohormones and Related Compounds-A Comprehensive Treatise*, Volume II, Letham, D.S., Goodwin, P.B. & Higgins, T.J.V. (eds.). pp 495-536. Elsevier/Northholland Biomedical Press, Amsterdam.

Yamaguchi-Shinozaki, K., Mundy, J. & Chua, N.-H. (1989) *Plant Mol. Biol.* **14,** 29-39.

Zeevaart, J.A.D. (1980) *Plant Physiol.* **66,** 672-678.

Glossary

The following terms are defined in a general sense. Specific details are avoided to allow the general reader and the developing student to become literate in the language (and jargon) underlying molecular biology and genetics.

abscisic acid: a plant hormone commonly associated with stress responses, but also with leaf abscission

AFLP: amplification fragment length polymorphism. A variant DNA amplification product of different size produced by DAF, PCR, or RAPD.

Agrobacterium rhizogenes:: bacterium causative for the hairy root disease, a form of organized tumor similar to the crown gall disease. Tumors also contain bacterial DNA transferred and covalently integrated in the plant genome.

Agrobacterium tumefaciens:: bacterium causative for the crown gall disease. capable of tumor induction in plants. This is achieved through the transfer of a region of DNA (T-DNA) which codes for the synthesis of plant growth regulators.

allele: one of the forms of a gene

aneuploid: chromosomal number is more or less than the diploid number by sets lesser than the haploid number (eg. a humans with 47 or 45 chromosomes are aneuploids.

antibiotic: substance which acts to destroy or inhibit the growth of a microbe (eg. bacteria or fungi).

antibody: common name for an immunoglobulin protein molecule which reacts with a specific antigen.

antigen: foreign molecule recognized by the immune response system of an animal.

ATP: adenosine triphosphate. The general energy donor molecule in all organisms.

autoradiography: method used to detect radioactive substances by the property to darken film superimposed on the compounds. Can be used on whole organisms or molecules separated by molecular methods such as electrophoresis.

auxin: plant hormone involved in cell elongation and growth (eg. indole acetic acid, 2,4-D).

bacteroid: the nitrogen-fixing form of a symbiosome-contained *Rhizobium* or *Bradyrhizobium* bacterium.

biolistics: process by which DNA molecules are propelled into a recipient cell using coated microprojectiles shot from a 'gene gun'. The method of propulsion may vary and ranges from electric discharge to helium blast.

biotechnology: the combination of biochemistry, genetics, microbiology, and engineering to develop products and organisms of commercial value.

Bradyrhizobium: bacteria able to form nodules and fix nitrogen in association with some legumes such as cowpea, soybean and peanuts.

cDNA: complementary DNA; DNA made by reverse transcriptase enzyme from RNA.

centimorgan: cM; unit of recombination equal to 1 percent recombination.

centromere: chromosomal region functioning as the spindle attachment region to allow chromosome and/or chromatid separation during mitosis and meiosis.

Chargaff's rules: stipulate that since in double stranded DNA the amount of adenine (A) equals that of thymine (T) and guanine (G) that of cytidine (C). Accordingly A binds to T and G to C by hydrogen bonds, giving the DNA molecule the properties needed for replication and information storage.

chloroplast: site of photosynthesis in eukaryotes. Contains circular DNA.

chromatin: complex form of eukaryotic nuclear material at the times between cellular divisions.

chromosome walking: strategy of chromosome analysis in which cloned chromosomal segments are used to isolate the neighboring DNA fragments.

chromosome: organized structure made up of DNA and proteins. Visible through light microscopy during cell division (cf. mitosis and meiosis).

clone: based on the Greek word 'klon' meaning 'twig'. A method of vegetative reproduction of an organism. Commonly used in horticulture as cuttings or drafts. Resulting organisms are defined as a clone meaning they were derived from the same original source organism. Commonly clones are presumed to be genetically identical. This may not be the case because of further genetic change after the original duplications. In modern genetic jargon, clone describes an isolated DNA sequenced ligated into a bacterial plasmid or virus, so that the sequence can be propagated indefinitely using microbiological means. Such cloned sequence can be used for sequencing, expression studies, or as probe.

codon: arrangement of three nucleotides in mRNA controlling the insertion of an amino acid into a polypeptide.

contig: contiguous fragment of DNA used in genome analysis.

cortex: bulk tissue of a plant root or legume nodule. Characterized by vacuolated cells and absence of mitotic divisions.

cyanobacterium: a prokaryote. Also called commonly blue-green alga. Capable of photosynthesis and sometimes of nitrogen fixation.

cytokinin: plant hormone involved in cell division and senescence (e.g. kinetin, zeatin).

DAF: DNA amplification fingerprinting. A method of general DNA amplification (cf. PCR) using a single primer of between 6 and 8 nucleotides in length. The primer is arbitrarily chosen and may generate as many as 80 amplification products, which can be resolved by a variety of methods, including polyacrylamide gel electrophoresis (PAGE) and silver staining, agarose electrophoresis, and automated analysis using either DNA sequencers (using fluorescent primers and laser detection) or capillary chromatography. Method was developed in 1990 by the University of Tennessee and is used to distinguish organisms (eg. cultivar of one crop species), gene mapping and diagnostics.

differentiation: process of cell and tissue specialization involving differential gene expression.

diploid: the normal somatic chromosome number of an organism (twice haploid).

DNA: deoxyribonucleic acid. The genetic molecule of most organisms, except some viruses. Double stranded polymer of nucleotides arranged along a deoxyribose and phosphate backbone. Structure was proposed by James Watson and Francis Crick in 1953.

electrophoresis: method used to separate protein or nucleic acid molecules in an electric field extending across a physical medium such as agarose gel or polyacrylamide. Derived from Greek: phoro, meaning I carry.

endodermis: cell layer separating cortex and pericycle in plant roots. Contains the Casperian Strip, which is important in diffusion resistance and turgor relationships in plants.

enzyme: a biological catalyst allowing the completion of biochemical reactions. Most enzymes are proteins although some RNA enzymes were recently discovered (see ribozyme).

epidermis: external cell layer of a plant.

Escherichia coli: also *E. coli*. A common gut bacterium used as a model genetic organism. *E. coli* has about 3000 genes and a genome of around 4 million basepairs.

ethidium bromide: chemical used to visualize DNA by fluorescence. Interpositions itself into the DNA groove and alters buoyancy.

ethylene: gaseous plant hormone involved in stress responses, fruit ripening as well as nodulation of legumes.

eukaryote: organism characterized by the presence of a nucleus. Also other organelles such as mitochondria and/or chloroplasts may be present in eukaryotes. Includes all plants, animals, green algae and fungi.

exon: expressed region of a gene. Transcribed and translated.

flavone: aromatic molecule (ie. contains a benzene ring as a core molecule) significant in the communication of legume plant to *Rhizobium* and *Bradyrhizobium*.

Frankia: an actinomycetes bacterium capable of forming nodules with some non-legumes such as Alnus and Casuarina (tree species).

gene: functional unit of inheritance. Usually a gene is defined as that region of DNA that controls the synthesis of a polypeptide.

genetic code: the conversion table which allows the interpretation of triplet codons to their matching amino acids.

genetic engineering: the directed genetic manipulation of an organism using recombinant DNA molecules not commonly found in nature.

genome: the entire set of hereditary molecules in an organism.

genotype: the genetic make-up of an organism which depending on the environment and other genes may be expressed to the phenotype.

geotropism: ability of a plant to sense the earth's gravitational field and to respnd positively towards the gravity.

gus: abbreviation for ß-glucuronidase, an enzyme that produces a blue colored pigment from a clor-less substrate. The gene coding for the enzyme is frequently used as a reporter gene in gene tranfer experiments.

haploid: the gametic chromosome number of an organism.

homogenotization: bacterial genetics procedure used to exchange genetic markers from a plasmid to the recipient linkage group. Also called marker exchange.

hormone: regulatory substance which acts at low concentrations (less than one micromolar) and at a distance from its site of synthesis. Controls metabolism and development.

intron: intervening sequence in genes. Transcribed but not translated.

isoflavone: aromatic signal substance involved in the nodulation of legumes.

karyotype: the pattern and shape of chromosomes of an organism.

kilobase: kb; one thousand basepairs of DNA.

lectin: protein molecule with no known function other than sugar-binding ability.

legume: plant family characterized by a pea-like flower morphology. Many but not all legumes are nodulated and form nitrogen-fixing symbioses with soil bacteria called *Rhizobium, Bradyrhizobium,* and *Azorhizobium.*

locus: the chromosomal position of a genetic condition as defined by a detectable phenotype.

Lotus japonicus: a legume used for forage. Because of its experimental advantages the plant has been conidered as a model for legume research.

map: the ordered arrangement of genes or molecular markers of an organism, indicating the position and distance between the markers and loci. Most maps are genetic maps based on the percentage recombination. Some maps are cytological maps based on the arrangement of chromosomal regions, while others are physical maps based on the amount of DNA between markers and loci.

megabase: Mb; one million basepairs of DNA.

meiosis: cell division in eukaryotes giving rise to gametes of different genetic make-up to the parental cell and to each other.

meristem: organized zone of mitotic division giving rise to cell clusters capable of further differentiation into new organ types.

messenger RNA: mRNA; product from DNA by transcription which serves as the information carrier for translation in proteins.

mitochondrion: organelle found in all eukaryotes. Site of respiration (ATP synthesis). Contains its own DNA.

mitosis: cell division in eukaryotes giving rise to two genetical identical progeny cells. Occurs frequently in meristems.

molecular marker: a molecular signpost used in eukaryotic gene isolation. Usually a RFLP probe or a primer site for DNA amplification.

mutant: an organism or gene with inheritable altered phenotype from the wild type

nitrogen fixation: process by which nitrogen gas is converted to ammonia. This process occurs frequently in bacterial induced nodules of legumes and results in an independence on fertilizer nitrogen. Also possible by the industrial Haber-Bosch process.

nod-box: DNA sequence found in front of several gene groupings involved in the nodulation ability of *Rhizobium* and *Bradyrhizobium*.

nodulation factor: lipo-oligosaccharide molecule (containing sugars and a fatty acid) synthesized by *Rhizobium* or *Bradyrhizobium* in response to a plant signal (usually a flavone or isoflavone) capable of inducing cell division and root hair curling in legume roots.

nodule: outgrowth from the roots (or stems in some cases) of legumes induced by bacteria or exogenous agents such as bacterial derived nodulation factors or auxin transport inhibitors.

nucleotide: component of DNA and RNA. Two nucleotides paired according to Chargaff's rules are one base pair.

oligonucleotide: a polymer of nucleotides usually 5 to 30 base pairs long.

organelle: membrane bound cellular compartment (eg. nucleus, chloroplast, mitochondrion, Golgi apparatus).

PCR: Polymerase Chain Reaction. A method for amplifying DNA of any organism using two specific oligonucleotide primers (about 15 base pairs in length) which flank the region of interest. The method was developed by CETUS Corporation in the mid-1980s and is of extreme value in diagnostics, forensics and general molecular biology (e. g. sequencing, probe preparation, genome mapping). Commercial rights are now owned by Hoffman-Roche.

peribacteroid unit: see symbiosome.

pericycle: cell layer surrounding the vascular bundle in plant roots. Gives rise to lateral roots as well as nodules in actinorhizal plants such as Alnus. Is also involved in legume nodule formation.

PFGE: see pulse field gel electrophoresis.

phenotype: the appearance of an organism taken as a genetic characteristic.

phosphorylation: biological process by which proteins are 'decorated' with phosphate groups derived from ATP. The process alters the biological activity of the protein whereby facilitating a form of physiological regulation.

phytoalexin: substance involved in the antimicrobial response of a plant.

plant growth regulator: broad class of chemicals which control the growth of plants. Many are also natural compounds found within plants, where they may act as hormones.

plasmid: circular, covalently closed DNA molecule commonly found in bacteria. Often used as a cloning vector in genetic engineering.

polyploid: more than diploid by multiples of the haploid number.

polysaccharide: polymer molecule of sugars (eg. starch, glycogen).

positional cloning: experimental approach used to locate and isolate gene sequences for which the gene product is not known. Instead the phenotype is mapped and large fragments are isolated in the region of informative molecular markers known to segregate closely with the gene of interest.

primer: short sequence of DNA (or RNA) used to initiate DNA replication.

probe: a known sequence of DNA (or RNA) used to detect homologous sequences in DNA or RNA after reassociation based on Chargaff's rules.

prokaryote: a bacterium. Characterized by the absense of major organelles such as the nucleus and plastids.

promoter: regulatory region of a gene involved in the control of RNA polymerase binding to the target gene.

protein: a polymer of amino acids usually with structural roles (such as keratin, the hair protein) or catalytic roles (see enzyme).

pulse field gel electrophoresis: PFGE; a variation on the electrophoresis procedure in as far that a computer flips the electric field in preset pulses in different directions and defined strengths. Method is used to isolate very large DNA molecules (greater than one megabase; million basepairs) including whole chromosomes of organisms such as yeast.

RAPD: Random amplified polymorphic DNA. DNA amplification method similar to DAF, as it uses single primers. Method is restricted to primers 9 nucleotides and larger, and amplification products are generally visualized using agarose electrophoresis and ethidium bromide fluorescence. Usually about 4 to 10 products are generated. Method is useful for mapping approaches.

rDNA: recombinant DNA; made by the joining of DNA fragments from different species using restriction endonuclease and cloning approaches.

receptor: protein molecule usually on the cell able to receive and interpret an external signal.

recombination: natural process of exchanging DNA fragments between different DNA molecules. Occurs in both prokaryotes and eukaryotes, but by slightly different processes. Eukaryotic recombination occurs predominantly during meiosis and gives rise to gametes of non-parental gene combinations.

restriction endonuclease: an enzyme which cuts (or restricts) DNA at specific sequences generating fragments (eg. *Eco*RI).

reverse transcriptase: enzyme able to synthesize DNA from RNA. Often found in tumor viruses.

RFLP: restriction fragment length polymorphism. DNA fragment difference generated by the action of a restriction endonuclease on the DNA of two or more organisms and detected usually by Southern hybridization using a radioactively labeled probe.

Rhizobium: bacteria able to form nodules with some legumes such as peas, alfalfa, and clovers.

ribosome: site of protein synthesis in prokaryotes and eukaryotes. Made up of two subunits, comprising three RNA molecules and about 50 proteins.

ribozyme: enzyme made entirely of RNA.

RNA: ribonucleic acid. Used as messenger RNA, transfer RNA and ribosomal RNA. Some RNA molecules are the genetic molecule of viruses (eg. tobacco mosaic virus and HIV) and viroids. Some RNA molecules may have enzymatic activity (see ribozymes).

rol **genes**: set of genes from *Agrobacterium rhizogenes* which cause the formation of transformed roots on a large variety of dicotyledonous plants. These genes seem to accentuate the normal plant hormone levels.

root hair: protruding from an epidermal cell. Grows by tip elongation.

Southern hybridization: also Southern blotting. Method employing gel separation of restricted DNA fragments, their blotting onto a membrane support, dissociation into single stranded DNA and hybridization (reassociation) with a labeled probe. Regions of homology are detected usually by autoradiography. Invented by Dr. Southern in 1975.

soybean: *Glycine max* (L) Merr. A tropical legume of widw agronomic application. Nodulates with *Bradyrhizobium japonicum* and *Rhizobium fredii*.

Stoffel fragment: truncated *Taq* polymerase without its exonuclease activity. Used in PCR and DAF. Named after the investigator Dr. Stoffel.

STS: sequence tagged site. Region of DNA on a chromosome used as a signpost for molecular gene mapping approaches.

supernodulation: ability of legume plants to form significantly higher nodule numbers than the normal. Found in several legumes such as pea and soybean.

symbiosis: mutually beneficial living together of two organisms of different species (eg. nodulation in legumes).

symbiosome: organelle structure found in nodules of legumes. Encases the symbiotic form of *Rhizobium* and *Bradyrhizobium* called the bacteroid. New term for 'peribacteroid unit (PBU)'.

Taq **polymerase**: thermostable DNA polymerase from *Thermus aquaticus*.

telomere: terminal region of chromosomes characterized by repeated DNA sequences.

Thermus aquaticus: thermophylic bacterium found in hot springs. Its DNA polymerase enzyme is thermostable and is used in the PCR, RAPD, and DAF.

transcription: the process by which DNA is copied into RNA. As the nucleic acid 'language' stays the same (see genetic code), the process is called transcription (cf. translation).

transformation: if used in a genetic sense, this term implies the transfer of a gene to a new cellular environment, coupled with the expression of that gene to alter the phenotype of the recipient cell. May be transient or stable as judged by inheritance.

translation: the process by which RNA is made into proteins. Occurs in ribosomes. Called translation because the nucleic acid 'language' based on a sequence of four 'letters' arranged in triplet codons is changed to a 20 component 'language' used in protein synthesis.

wild type: the normal form of an organism (ie. not mutant). Note: this is a noun (not hyphenated).

wild-type: the normal, non-mutant form of an organism or allele. Note: this is the adjective and is hyphenated.

YAC: yeast artificial chromosome. Used extensively in the cloning of large DNA molecules used in eukaryotic gene mapping.

DATE DUE